A WORLD BANK COUNTRY STUDY

Paraguay
Country Economic Memorandum

The World Bank
Washington, D.C.

Copyright © 1992
The International Bank for Reconstruction
and Development/THE WORLD BANK
1818 H Street, N.W.
Washington, D.C. 20433, U.S.A.

All rights reserved
Manufactured in the United States of America
First printing August 1992

World Bank Country Studies are among the many reports originally prepared for internal use as part of the continuing analysis by the Bank of the economic and related conditions of its developing member countries and of its dialogues with the governments. Some of the reports are published in this series with the least possible delay for the use of governments and the academic, business and financial, and development communities. The typescript of this paper therefore has not been prepared in accordance with the procedures appropriate to formal printed texts, and the World Bank accepts no responsibility for errors.

The World Bank does not guarantee the accuracy of the data included in this publication and accepts no responsibility whatsoever for any consequence of their use. Any maps that accompany the text have been prepared solely for the convenience of readers; the designations and presentation of material in them do not imply the expression of any opinion whatsoever on the part of the World Bank, its affiliates, or its Board or member countries concerning the legal status of any country, territory, city, or area or of the authorities thereof or concerning the delimitation of its boundaries or its national affiliation.

The material in this publication is copyrighted. Requests for permission to reproduce portions of it should be sent to the Office of the Publisher at the address shown in the copyright notice above. The World Bank encourages dissemination of its work and will normally give permission promptly and, when the reproduction is for noncommercial purposes, without asking a fee. Permission to copy portions for classroom use is granted through the Copyright Clearance Center, 27 Congress Street, Salem, Massachusetts 01970, U.S.A.

The complete backlist of publications from the World Bank is shown in the annual *Index of Publications*, which contains an alphabetical title list (with full ordering information) and indexes of subjects, authors, and countries and regions. The latest edition is available free of charge from the Distribution Unit, Office of the Publisher, Department F, The World Bank, 1818 H Street, N.W., Washington, D.C. 20433, U.S.A., or from Publications, The World Bank, 66, avenue d'Iéna, 75116 Paris, France.

ISSN: 0253-2123

Library of Congress Cataloging-in-Publication Data

Paraguay—country economic memorandum.
    p. cm.—(A World Bank country study)
  "Draws on the findings of missions that visited Paraguay in March
and November 1990"—P.
  ISBN 0-8213-2158-7
  1. Paraguay—Economic policy. 2. Paraguay—Economic
conditions—1954– I. International Bank for Reconstruction and
Development. II. Series.
HC222.P442 1992
330.9892—dc20   92-17097
                                                           CIP

PARAGUAY

COUNTRY DATA

Area	Population	Density
(in thousands sq. kms)	4.0 million (1989)	9.8 per sq. km
406.8	Rate of Growth 2.9 (1989)	

SOCIAL INDICATORS (1987)

Population Characteristics
- Crude Birth Rate (per 1,000)   35
- Crude Death Rate (per 1,000)   7
- Infant Mortality Rate (per 1,000 live births) 42

Health
- Population by Physician   1800
- Population per Hospital Bed   700

Access to Safe Water
Percent of Population - Urban   46
                     - Rural   10

Access to Electricity
Percent of Dwellings - Urban   52
                    - Rural   40

Nutrition
- Calorie Intake   2873
- Per Capita Protein Intake (grams per day)   81.0

Education
- Primary School Enrollment (%)   101.0

PER CAPITA GROSS NATIONAL PRODUCT IN 1989: US$1010.4

Gross National Product in 1989	US$ Million	% of GNP	Annual Growth Rates (%, Constant Prices)			
			1977-81	1982-87	1988	1989
GNP at Market Prices	4041.4	100.0	11.1	0.7	7.0	5.4
Gross Domestic Investment	898.2	22.2	20.2	-4.8	3.7	-1.2
Gross National Savings	802.5	19.9				
Current Account Balance	95.7	2.4				
Exports of Goods & NFS	1466.6	36.3	1.6	11.0	8.4	10.7
Imports of Goods & NFS	1464.9	36.2	9.8	-3.6	4.2	-13.6

OUTPUT, LABOR FORCE AND PRODUCTIVITY IN 1989

	Value Added		Labor Force		Value Added/Per Worker	
	US$ Million	Percent	Thousands	Percent	US$ Million	Percent
Agriculture	809.5	20.0	570.0	36.8	1420.0	54.4
Industry	1212.3	30.0	315.0	20.0	3848.0	147.4
Services	2019.6	50.0	663.0	42.8	3046.0	116.7
Total/Average	4041.4	100.0	1548.0	100.0	2610.0	100.0

GOVERNMENT FINANCE

	Cons. Public Sector			Central Government		
	US$ Million	% of GDP		US$ Million	% of GDP	
	1989	1989	1984	1989	1989	1985
Current Receipts	635	15.4	13.1	408	9.9	7.9
Current Expenditures	446	11.8	11.1	309	7.5	6.7
Current Balance	148	3.6	2.0	99	2.4	1.2
Capital Expenditures	219	5.3	8.4	70	1.7	2.5

Money, Credit and Prices

	1986	1987	1988	1989
	(billions of guaranies outstanding at end of period)			
Money and Quasi Money a/	274.7	365.6	430.6	609.4
Bank Credit to Public Sector b/	42.8	46.2	10.7	-102.0
Bank Credit to Private Sector	166.8	211.3	276.8	384.2
Money and Quasi Money as % of GDP	15.0	14.7	13.0	13.2
General Price Index, Average (1982 = 100)	239.3	311.9	390.4	512.7
Annual Percentage Changes in:				
General Price Index	31.5	30.3	25.1	31.3
Bank Credit to Public Sector	21.9	7.9	-76.8	-1053.3
Bank Credit to Private Sector	35.8	26.7	31.0	38.8

Balance of Payments

	1987	1988	1989
	(US$ millions)		
Exports of Goods, NFS	872.4	1190.8	1466.6
Imports of Goods, NFS	1238.2	1484.0	1464.0
Resource Gap (deficit -)	-365.8	-293.2	2.6
Interest Payments	-150.4	-154.6	-98.5
Other Factor Payments (net -)	28.4	70.7	63.0
Net Current Transfers	27.0	35.3	7.2
Balance on Current Account	-460.8	-341.8	-25.7
Direct Investment	8.5	6.2	21.0
Public M&LT Loans (net)	88.7	-21.9	125.3
Disbursements	305.2	220.8	654.5
Amortization	216.4	242.7	529.2
Other Capital (Net)	292.1	168.5	86.1
Increase in Reserves (-)	55.9	-135.3	206.7
Arrears	127.4	53.7	0.0
Gross Reserves (end Year)	483.3	348.0	554.7

Merchandise Exports (1989)

	US$ million	Percent
Cotton	351.6	30.2
Soybeans	467.2	40.1
Timber	26.6	2.3
Livestock	136.0	11.7
Others	183.0	15.7
Total	1,164.0	100.0

M&LT External Debt, 12/31/89

	US$ million
Public Debt inc. Guaranteed	2604.7
Non-Guaranteed Private Debt	26.7
Total Outstanding & Disbursed	2631.4

Debt Service Ratio for 1989

	Percent
M&LT Public Debt inc. Guaranteed	21.1

IBRD Lending, 12/31/89

	US$ million
Outstanding & Disbursed	336.5
Undisbursed	48.0
Outstanding inc. Undisbursed	384.5

Rate of Exchange c/

1980: US$1.00 = G126.0
G1.00 = US$.0079
1989: US$1.00 = G1120.0
G1.00 = US$.00089

---

a/ Private Sector only
b/ Net of Deposits
c/ Average Annual Rates

CURRENCY EQUIVALENTS

Currency Unit = Guarani (G) (As of April 20, 1992)
US$1.00 = 1440 G$
G$ 1.00 = US$0.0006944

FISCAL YEAR

January 1 to December 31

GLOSSARY OF ABBREVIATIONS AND ACRONYMS

ACEPAR	Steel Company
ANAC	Civil Airports Administration
ANDE	Power Comany
ANNP	Ports Administration
ANTELCO	Telecommunications Company
APAL	Alcohol Company
BCP	Central Bank of Paraguay
BNF	National Development Bank
CORPOSANA	Water and Sewage Company
FCCAL	Railroad Company
FLOMERES	Shipping Transport Company
IBR	Rural Welfare Institute
IDB	Inter-American Development Bank
INC	Cement Company
IPS	Social Security Institute
IPVU	Housing Institute
LAP	Paraguayan International Airline
LATN	National Transport Airline
MEC	Ministry of Education
MH	Ministry of Finance
MIC	Ministry of Industry and Commerce
MOPC	Ministry of Public Works and Communications
MPSB	Ministry of Health and Social Welfare
NBT	Workers National Bank
PETROPAR	Oil Importing Company
SIDEPAR	Steel Holding Company
STP	Planning Secretariat
UNA	National University
WB	World Bank

ACKNOWLEDGEMENTS

This report draws on the findings of missions that visited Paraguay in March and November 1990. The first mission was led by J. Garcia-Mujica and integrated by M. Cortes (Trade), C. Corti and Y. Boray (Public Enterprises), and M. Preece (Statistical Annex). Contributions also were made by A. Fleming (Finance), J. Joyce (Agriculture), and G. Unda and W. Matthey (Transport) based on separate missions. The report also summarizes and integrates a number of studies on the Paraguayan economy indicated in the References section. The report was written by J. Garcia-Mujica.

TABLE OF CONTENTS

Page Nos.

EXECUTIVE SUMMARY AND RECOMMENDATIONS ix

CHAPTER I: GROWTH IN THE LAST HALF CENTURY 1

    Instability and "Stagflation" in the Forties and Fifties 1
    The Sixties: Establishing the Basis for Faster Growth 3
    Itaipu: The Basis for Accelerated Growth in the 1970s 5
    The 1980s: Macroeconomic Instability and Stagnation 8
    Credit and Monetary Policies 20

CHAPTER II: SECTORAL CONSTRAINTS AND OPPORTUNITIES FOR GROWTH . 24

    A. Agriculture 24
        Productivity 24
        Agriculture Taxation 25
        Environmental Considerations 25
        Future Agricultural Development Strategy 26

    B. Population, Colonization and Poverty 28
        Official Emphasis on Foreign Immigration Until the 1940s 28
        Colonization to Fight Poverty After 1940 30
        Difficulties Ahead 31

    C. Transport 31
        Main Agreements with Brazil 32
        Evolution of the Road Network 33
        Challenges Ahead 34

    D. Itaipu and Yacireta: Electricity as an Export 35

    E. Informal Trade 39

CHAPTER III: POLICY REFORMS TO STIMULATE GROWTH 41

    A. Financial Sector Reform 42

    B. Trade Policy Reform 49
        The Present Customs Law 49
        Revenue Collection 50
        Intended Effect of the Customs Law 53
        Impact of Customs Duties by Sector 54
        Informal Imports 56
        The Challenges of MERCOSUR 57
        Export Incentive Policies 57
        Summary 58

C. Public Enterprise Reform	58
ANDE	62
ANTELCO	64
CORPOSANA	65
FLOMERES	66
INC	68
LAP	69
ACEPAR	71
PETROPAR	72
Summary	74
D. Tax Reform	75
The Present Tax System	77
Ongoing Administrative Reforms	79
The Present Tax Reform	79
Reflections on a Tax Reform	81

CHAPTER IV: ECONOMIC PROSPECTS 83

Summary Background 83
Low Case: Retreating from the Changes Made 85
High Case: Fast Implementing Missing Changes 88
Intermediate Case: Slower Progress but No Retrogression 90
Conclusions 91

REFERENCES 92

ANNEX I - HISTORICAL BACKGROUND 95

ANNEX II - SUMMARY OF SCENARIOS LOW, MEDIUM AND HIGH 102

STATISTICAL APPENDIX 124

COUNTRY MAP

EXECUTIVE SUMMARY AND RECOMMENDATIONS

Overview

Agriculture has been Paraguay's main source of growth since colonial times. After two devastating wars, in 1865-70 and 1932-35, the wealth of the land rescued the country from disaster. More recently, between 1940 and the mid-1960s, the fortunes of agricultural exports were a major determinant of the rhythm of growth. New sources of growth developed in the 1960s: commerce and tourism, and construction. In the 1970s, growth accelerated sharply, with the rapid growth of agriculture based on the expansion of the agricultural frontier and the construction of Itaipu dam. Itaipú dominated economic performance both in the 1970s, when the civil works were the major factor in aggregate growth, and in the 1980s, when the economy had to face the consequences of the end of the massive construction expenditures. In the future, sales of surplus hydro-power from Itaipu will benefit Paraguay substantially, as will sales from Yacireta when it is completed.

Paraguay now stands at a critical but highly promising crossroads. An authoritarian regime has been deposed and the new government has taken a number of strong actions to restore stability, including improving public finances, bringing down inflation, unifying and liberalizing the exchange rate, freeing many interest rates, and initiating a tax reform. Yet Paraguay's history contains numerous examples of long, strong and pervasive authoritarian regimes followed by periods of lingering and severe political instability and economic decline. The challenge for the Government is to avoid such economic deterioration. The current Government has an almost unique opportunity to develop a working democracy in an environment of financial stability and strong and sustainable economic growth. Along with sound macroeconomic management, a number of structural changes would contribute to these goals, including shifts in agricultural policy to promote a more land-intensive pattern of development; continued investment in and maintenance of transport infrastructure; changes in financial sector regulation to reduce the artificial barriers between institutions and improve financial supervision; reduction and unification of tariffs to bring the Customs Law into line with actual practice and to prepare for entry into the MERCOSUR. Some of these actions may require additional resources. To achieve this, it will be necessary to improve public enterprise management including privatizations and joint ventures, and to reform taxes by lowering rates to collectible levels, widening the base, and improving collections from agriculture.[1]

---

[1] Social sector needs--water and sewerage, education, health and nutrition--also are urgent. An upcoming report, Paraguay, Public Expenditure Review: The Social Sectors suggests that although Paraguay's social indicators are surprisingly good in light of its per capita income and low public expenditures, a number of problems remain to be addressed. Particularly important is to improve the efficiency of public expenditures in these areas. If this is not achieved, merely increasing expenditure, in an attempt to ameliorate the problems, would have no major impact on social indicators and resources would be wasted.

The Changing Sources of Paraguay's Growth

i.   Traditionally, external demand for agricultural products has been an important factor in Paraguay's prosperity. The 1940s and 1950s were no exception. Economic growth was high during World War II--demand for Paraguayan exports was high due to war-related needs among the Allies. After 1945, external demand dropped and the economy stagnated. Growth resumed in the second half of the 1950s, although at modest rates. From the late 1930s through the 1950s, inflation was high and variable, reducing the incentives to save and invest.

ii.   The 1960s were years of financial and political stability that witnessed a change in the sources of growth. Agriculture remained the main engine of growth in the first half of the 1960s, but was not a major factor in the second half, despite the beginnings of the expansion of the agriculture frontier during the decade. After 1965, the main sources of growth became commerce and construction. Paraguay's essentially open frontiers allowed it to take advantage of the protectionist policies of its neighbors and become a commercial and tourist center. Informal trade expanded rapidly, especially after road links with Brazil were improved. The internal road network also was expanded substantially. The construction of the country's first hydroelectric plant, along with the signing of agreements that paved the way for the construction of the largest hydroelectric plant in the world, Itaipu, foreshadowed enormous growth in the energy sector.

iii.   Growth accelerated dramatically in the 1970s with the construction of Itaipu. As in the second half of the 1960s, construction and commerce were the most dynamic sectors of the economy in the 1970s, but this time agriculture's performance also was outstanding, helped by the expansion of the frontier begun in the 1960s. Investment increased sharply, from 15 percent of GDP in 1970 to 29 percent in 1981. (Investment figures exclude construction of the Itaipu works, which is treated as a foreign operation from the standpoint of Paraguay's national accounts and balance of payments). Inflation accelerated, largely in parallel to the rise in international rates. The balance of payments showed large surpluses, reflecting the large inflows of foreign exchange related to the construction of Itaipu and the control of domestic credit.

The Macroeconomic Instability of the 1980s

iv.   After the completion of the main works at Itaipú in 1981, the Paraguayan economy entered a deep recession that lasted well into the mid 1980s. GDP declined for two consecutive years, leaving production 4 percent lower in 1983 than in 1981. Growth resumed in 1984, but the average rate over 1984-86 was well below the growth of population (2 and 3 percent per year correspondingly). The recovery speeded up in 1987 and 1988, but this was not enough to offset the losses in per capita income arising from the 1982-86 recession. Social unrest burst into a revolution in February 1989, which toppled a regime that had lasted for 35 years

v.   The sectors that suffered most in the post-Itaipu era--construction, and commerce and finance--were those that had grown most rapidly in the 1970s boom. Stagnant domestic demand, linked to the fall in foreign exchange receipts from Itaipu, was the main factor in the slow-down As the private sector mistook transitory income increases stemming from the construction of Itaipú with permanent increases in wealth, many of the investments that had been made in the boom of the 1970s were ill-designed for the post-Itaipu period, and proved imprudent. Data suggest that exports continued to grow in the 1980s but at a slower rate than in the 1970s. However, the poor quality of the statistics--Paraguay's porous borders mean that large volumes of trade appear and disappear in the official statistics depending on policies in the neighboring countries--also suggest caution is needed in drawing this conclusion or even in estimating the severity of the downturn of 1982-86.

vi. The Government tried to avert the post-Itaipu recession by embarking on an investment and spending program. This only piled a financial disequilibrium on top of the recession. The economy still stagnated, and inflation accelerated. To finance its increased spending, the Government sharply increased its external borrowing. In addition, credit to the public sector (including the Central Bank's deficit) also increased; this contributed to rising inflation and, despite the foreign borrowings, loss of the abundant international reserves that had been built up in the 1970s. Ceilings on interest rates provoked a massive disintermediation of quasi-monetary savings that contributed to the inflation and the loss of reserves. In a vain attempt to stem the balance of payments problem, the Government instituted a complex system of multiple exchange rates. This not only failed to stop the loss of reserves, it distorted the economy and encouraged corruption, the more so as the domestic price level increased. The principal legal beneficiary of the multiple exchange rate system was the nonfinancial public sector. Although official figures suggest the nonfinancial public sector deficit remained roughly constant throughout the 1980s, the explicit and implicit annual subsidies from the Central Bank and exporters to the nonfinancial public sector reached the equivalent of almost 5 percent of GDP annually in 1985-88.

vii. As a result of the sharp increase in external borrowing, external debt rose substantially in the 1980s. Medium- and long-term debt (including arrears) rose from 15 percent of GDP in 1981, to 62 percent of in 1987. Moreover, the external funds were largely used on unprofitable projects, which added to the macroeconomic instability and eventually led to the suspension of disbursements from several creditors. In the mid-1980s, the Government stopped servicing much of its international debt. Arrears to commercial banks and bilateral creditors have accumulated since then.

viii. The new Authorities that took charge after February 1989 have made substantial improvements in policy management. The government deficit was reduced, public enterprise finances were tightened significantly, the multiple exchange rates were unified and the exchange market liberalized, interest rates were freed, and a tax reform was begun. Although inflation increased from 15 percent in early 1989 to 44 percent in 1990, it was cut to a less than 15 percent annual rate in 1991. The balance of payments has improved, although Paraguay continues to accumulate external arrears. However, a number of policy reforms are still needed in order to strengthen the sectors that will form the basis of Paraguay's growth in the 1990s and improve prospects for high and sustainable growth.

Sectoral Constraints and Opportunities for Growth

ix. **Agriculture** still provides the basis upon which much of the country's economic activity rests. It employs 40 percent of the labor force, supplies raw materials used in manufacturing, and was the only significant source of foreign exchange earnings until the construction of Itaipu. The potential of the agriculture sector, in terms of underutilized resources, available technology and markets, should mean that the sector can continue to play an important role in economic growth and development. The main issues in achieving rapid and sustainable agricultural growth are related to: increasing productivity, further expanding the agriculture frontier, ensuring careful management of renewable resources, and removing the distortions imposed by the tax system.

x.	Improving agricultural productivity to reach the higher yields obtained under controlled conditions, or even in neighboring countries, is always a difficult task. An important start could be made by encouraging more intensive use of resources via sustainable agriculture practices, rather than further expansion of the agriculture frontier. Such an approach would begin by improving land information systems, preparing new cadastres, setting rules on allowed uses of land and penalties for breaking them, and abolishing remaining regulatory the impediments to imports and local distribution of fertilizers.

xi.	Improvements in the management of renewable resources also would contribute to higher productivity. In particular, deforestation is an increasingly critical issue, with forest in many areas being removed with no concern for the land's long-term productivity. Current policy gives conflicting signals for forest management and subsidized credit has contributed to deforestation. Paraguay needs to define and enforce a consistent, environmentally-sound set of guidelines.

xii.	Changes in the tax policy toward agriculture also would encourage more intensive land use. The existing estate tax often goes uncollected and property assessments do not reflect market values. At a later stage, the land tax could be refined into one more closely related to the capability of the soil, making it production-positive by allowing deductions linked to capital investments in the land.

xiii.	**Colonization and the Agricultural Land Problem**. Paraguay was and remains one of the least densely populated countries in Latin America. The country's population was decimated after the Triple Alliance War in 1865-70; the Chaco War (1932-35) was a further setback. Underpopulation typified the country into the late 1950s, so the Government encouraged immigration. Foreign immigrant colonies were officially sponsored and successfully established between the 1880s and the 1940s. Spontaneous Brazilian immigration was significant during the 1960s and 1970s.

xiv.	The official focus of the colonization program changed after the 1940s, from settling the still-empty regions to poverty alleviation. New settlements were established in the country's rich eastern-most regions, where spontaneous colonization had already started by the 1950s. The Instituto de Bienestar Rural (IBR) was founded with this purpose in 1963. Between its inception and 1982, officially sponsored and private settlements covered 6.5 million ha. on 93,000 lots benefiting more than half a million people, about one third of the rural population in the early 1980s. This temporarily eased social tensions in the 1970s.

xv.	Despite these efforts, the agrarian land problem and its links to the minifundia/poverty problem remain urgent. Land occupations and evictions are becoming more common. IBR could help alleviate these problems by improving its system of granting land titles and providing landholders with more secure titles. Further land distribution may be another option, but it would involve larger land purchases from the private sector, with greater financial costs. Improvements in IBR's collections from land transfers would be critical to limit the cost of the program. But even this would not be enough. Perhaps more importantly, reducing poverty will depend on a shift in the pattern of agriculture growth to more intensive land use and an efficient increase in social expenditures in rural areas (see footnote 1).

xvi. **Transport**. Lack of transport infrastructure has been a major bottleneck to the country's efficient use of resources. Until the sixties, many areas were inaccessible to wheeled traffic or could be reached only under the most favorable weather conditions. Since the late thirties the Government has been committed to reducing internal and external dependence on riverways by expanding the road network. The internal road network multiplied four times between 1940 and 1955, almost doubled between 1955 and 1960, and nearly doubled again in the following five years. The expansion of the domestic road network slowed somewhat after 1965, but road mileage continues to grow faster than GDP. Agreements with Brazil, reached in the 1940s and 1950s, formed the basis of expanded international communications by land, especially after the construction of the bridge over the Parana River at Ciudad del Este was finished in 1965.

xvii. The results of the expansion in the road network were impressive. The country became much less dependent on river transport, the expanded agricultural frontier could be serviced, and trade with fast-developing Brazilian regions (Parana and Matto Grosso) was greatly facilitated. Nonetheless, the transport sector faces important problems. Financing of improvements in river ports and channels could become an important issue if such improvements, plus the prospective lower costs of privatized ports in Argentina and Uruguay, reduce river shipping costs of bulk exports such as cotton and soybeans below overland shipping costs through Brazil. A decision must be taken about the future of the railway, which has deteriorated to the point where it has almost no commercial value. Regarding roads, with limited resources the Government faces difficult choices: improving existing roads, maintaining the network, or building new penetration roads.

xviii. **Energy as a Source of Foreign Exchange**. Paraguay has abundant hydroelectric capacity; sales of surplus electricity to neighboring countries are projected to be a major export in the 1990s. Paraguay began to tap its hydroelectric resources when Itaipú was built; Yacireta will bring further benefits when it is finished. Paraguay's revenues from Itaipu will stem mainly from two sources, royalties charged for the use of Paraguayan waters in the Parana River, and "sales" of surplus electricity to Brazil (the surplus electricity assigned to Paraguay is transferred back to the Itaipu Binational Corporation, which sells it to Brazil at a price agreed in the Itaipu Treaty and modified in later referral notes). However, receipts have been temporarily reduced, first by delays in dam construction; second, because Paraguay agreed to lend back to Itaipu a fraction of the income to which it was entitled, and third because of Brazil's financial difficulties (Brazil did not fully pay Itaipu and therefore Itaipu could not pay Paraguay). Beginning in 1991, Itaipu will be operating at full capacity. Beginning in 1992, Paraguay is projected to receive nearly US$200 million annually from royalties and sales of surplus electricity, provided that Brazil can pay on time. This income is equivalent to more than 3 percent of Paraguay's GDP.

xix. **Informal Trade**. Informal trade appears to be significant in Paraguay, but few reliable estimates of the sector's size are available. Unregistered import and export transactions cause concern because they don't contribute their fair share to Government finances. But they also have had a positive effect, namely, setting a limit on the distortions in relative prices that in other Latin American countries became so pervasive. Informal trade was fostered by the construction of the bridge over the Parana River at Ciudad del Este and the related improvements in road connections with Brazil. To a large extent, informal trade reflects the protectionist policies prevailing in neighboring countries. Thus, it may suffer if Brazil and Argentina open up their economies to international competition, or if MERCOSUR is successfully implemented.

New Sources of Growth and Associated Policy Reform

xx.   **The Financial Sector.** Several problems need urgent attention in the financial sector. First, the health of the financial system, which weakened after the financial crisis in 1982, must be carefully evaluated and nurtured, to define a safe path that will encourage financial savings. Second, homogeneous regulations and efficient supervision of financial institutions are needed to improve the financial sector's mobilization and allocation of resources. Third, the roles of the official banks need to be changed and some official banks may need to be closed.

xxi.   The government has liberalized interest rates on deposits and loans, except the discount rate. Full interest rate liberalization might have created difficulties after the 1982 crisis, when, due to a strong devaluation that affected many creditors, bank portfolios deteriorated dangerously. Currently, however, although some private banks are in poor shape, most private banks seem reasonably capitalized and can respond to the pressures of the market. The report recommends maintaining the regime of free rates and placing greater reliance on market forces in determining the Central Bank's rediscount rates.

xxii.   Bank supervision concentrates too much on the supervision of compliance with compulsory investments, tax laws, and reserve requirements and too little prudential supervision (for example, evaluating portfolio quality and capital adequacy), to the detriment of the system's health. Another problem is differential regulation in the financial system, which has discriminated against private banks, produced an artificially segmented financial system, and overloaded the Superintendency of Banks. The Government is revising the Central Bank Law and the banking laws. The report recommends that these laws and/or their complementary legal provisions (i) redefine the role of the Superintendency of Banks and its relation with the Central Bank, so as to encourage the proper kind of supervision, and (ii) reduce artificial distinctions between financial institutions.

xxiii. The main official banks (in addition to the Central Bank) are the National Development Bank (BNF), that lends to small farmers, the Livestock Fund (FG), that lends to large ranchers, the National Housing Bank (BNV), that supervises savings and loans institutions (serving credit needs in the housing sector), and the National Workers Bank (BNT), that supposedly lends to workers. Official banks lost vast sums by borrowing in foreign currency and lending locally at fixed interest rates. Recapitalizing the official banks would require a large commitment from the government; hence, it is worth examining their roles in some detail at this point. BNF probably deserves support if it can fulfill its role effectively. However, because reaching small farmers is a socially desirable but expensive endeavor, BNF is unlikely to generate profits. Alternative solutions need to be examined, such as greater reliance on private banks, to keep subsidies clear and transparent and ensure timely collections. Regarding FG, there is no reason for it to operate with losses, as large ranchers should be able to pay market interest rates for the credit they receive and should finance the extension services they use. Thus, the report recommends that FG's role be undertaken by the private sector and advises against transforming FG into another public bank. The report suggests that there is little need or justification for having BNV in Paraguay. The BNT should operate like any other commercial bank, if it cannot, it should be closed within a reasonable time-frame.

xxiv. **Trade Policy Reform**. Trade-related distortions are not important in Paraguay now. However, they easily could become sizable in the future unless action is taken to bring the tariff code in line with actual practice. In practice, and in relation to other countries, non-tariff and tariff barriers are low. However, the existing Customs Law and some related taxes contain a heavy protectionist bias. Custom duties (including surcharges) range from 3 percent to 86 percent, and tend to be higher for traditional manufacturing sectors. If applied, these duties would imply high effective protection. Actual protection is limited, however, because the Government has implemented several simple special tariff regimes (border countries with a 10 percent flat rate; tourism with a 7 percent flat rate; and whisky and tobacco with a 8 percent flat rate), and because unregistered trade is widespread. The high tariffs have not generated large fiscal revenues and also have failed to provide protection to domestic production; Paraguay is *de facto* a free-trade economy with low actual protection. To remove the possibility that the Customs Law might be applied, to ensure the continued benefits of low protection, and to allow the country to keep pace with reductions in tariffs taking place in its MERCOSUR partners, the report recommends bringing the Customs Law into harmony with actual practice by enacting a flat 10 percent import tariff.

xxv. **Public Enterprises**. The financial condition of public enterprises remains weak, despite some improvements in 1989 and in 1990. Public enterprise deficits still exceed the levels at the beginning of the 1980s, and, as with the government, public enterprises are accruing arrears on much of their external debt. Public enterprise performance deteriorated in the 1980s because of poor control polices, overstaffing, lack of managerial accountability, and, until recently, poor pricing policies.

xxvi. Major improvements in financial management took place in 1989-90 that partly offset the elimination of the large subsidy on foreign exchange that the public enterprises enjoyed. Cuts in some investments also occurred, linked to the postponement of some works and the completion of others (including cement and steel plants of doubtful value). Nonetheless, the public enterprises' overall deficit increased from about 3.4 percent of GDP in 1988 to an estimated 4.4 percent in 1990. (This comparison overstates the increase in the deficit because the 1988 figure is on a cash basis and the 1990 figure on an accrual basis, but the deficit on an accrual basis in 1988 would have been much larger than on a cash basis because of the prevailing subsidy on foreign exchange.) Moreover, the public enterprises' current saving as a share of GDP also declined sharply in 1989. The large increases in tariffs in 1990, particularly on petroleum prices and water, generated new resources and increased current saving to nearly the 1988 level, but were not enough to reduce to deficit to 1988 levels.

xxvii. The report concludes that further improvements in productivity are needed. Plans to reduce overemployment are in place and must be implemented. Overdimensioned investments were an important problem in the recent past. The concern now is to avoid repeating such mistakes. The Authorities are planning to improve control and accountability in public enterprise management, and audit processes are in progress in several enterprises; these actions should be strengthened. Price and tariff distortions were an issue in the 1980s, but actions taken after February 1989 have removed some of the most important such distortions. The main price distortions remaining are in potable water (despite the recent price increase) and, especially, sewage. Although most public sector tariffs are reasonable, a method must be developed to ensure that tariffs are regularly adjusted and, at the same time, pressure is put on enterprises to avoid their covering up inefficiencies through higher relative prices. Privatization is one method for improving the delivery of services at lower cost to the State. A number of state enterprises could be privatized or operated as joint ventures and the Government is beginning to study these options.

xxviii. **Tax Reform**. Although tax revenues increased sharply in 1989, Paraguay's tax system remains antiquated and inefficient. The tax system is based on: multiple specific taxes with widespread exemptions instead of a few generalized ad-valorem ones, outdated bases for property taxes, and payments made with long delays and no effective penalties. As a result of inappropriate rates and bases, tax collection did not keep pace with domestic inflation and growth in the 1980s. In particular, because of lower revenues from import taxes, linked to the inefficient system of multiple exchange rates prevailing until 1989, tax revenues dropped two percentage points of GDP. The cadastre for taxing properties is obsolete and needs updating; in fact, many rural estates have never been assessed. Evasion is widespread because of loopholes (only some of them legal) and the lack of effective fines; low interest penalties encourage late tax payments. Hence, a better sanctioning system must be designed.

xxix. Tax administration is also weak. Until recently, there were several independent tax units with similar tasks and organizations, an institutional arrangement that fostered confusion and waste. Recently, the Finance Ministry has made substantial improvements in its organization and in tax administration in particular. The report applauds these changes, since efficiency will be enhanced by eliminating duplications.

xxx. A significant and overdue tax reform is currently under preparation. The Finance Ministry has submitted a tax reform package to Congress involving a radical simplification of the system: few taxes with ad-valorem rates on wide bases, a value-added tax replacing the present sales tax, reassessing property values, and a more efficient system of fines. The report suggests higher penalties and cuts in tax rates, to levels that are collectible, in exchange for radical increases in tax bases (if possible, no exceptions).

Prospects

xxxi. Paraguay has made substantial progress towards regaining macroeconomic stability. Many painful aspects of adjustment already have been completed successfully (i.e., freeing the exchange rate, liberalizing most interest rates and strengthening public finances). However, additional efforts are necessary. If the issues discussed above can be resolved, then prospects for high growth rates and macroeconomic stability are excellent.

xxxii. The report develops three alternative scenarios to illustrate the importance of completing the adjustment process already under way and securing international cooperation. A Low Case illustrates the negative consequences of delaying indefinitely the adjustments recommended in the report and reversing progress made in exchange rate policy, interest rate policy and tax administration. Under this scenario, inflation mounts, distortions increase, and capital outflow occurs, making this an unviable alternative. The High Case illustrates the effects of rapidly implementing the needed changes, namely the reforms suggested in the financial sector, the simplification of the tariff code, the efficient increase in infrastructure and social sector spending, the improvements in efficiency in public enteprises, the privatization of selected public enterprises, and the successful enactment of the tax reform--the latter two also are needed to generate the increase in public sector resources for high priority expenditures. Results are encouraging: GDP growth accelerates and remains high, inflation drops rapidly to international levels, the balance of payments remains strong, and creditworthiness is strengthened. The third scenario is an Intermediate Case which reflects the likely advance of reforms; a slower pace than in the High Case but no retrogression. Results are favorable, but of course somewhat less so than in the High Case. Inflation drops to single digit figures, growth reaches nearly 5 percent per year, and external accounts are nearly as strong as in the High Case. Regarding foreign financing requirements, in the Low Case, due to poor domestic policies, voluntary disbursements will not be available; the country resorts to the unattractive and unproductive option of accumulating arrears. In the High and Intermediate Cases, less external credit actually is needed than in the Low Case because the improved policy framework stimulates domestic saving and foreign investment and reduces capital flight. External credit is expected to be provided voluntarily, mainly by bilateral and multilateral lenders. Resolution of Paraguay's arrears problem, coordinated support from foreign lenders, and good project preparation are the key constraints facing the country in the external area. Both the Intermediate and High Cases assume the arrears problem is resolved in 1992.

CHAPTER I: GROWTH IN THE LAST HALF CENTURY

1.1　　　　Paraguay's growth in the last 50 years occurred in four distinct phases.[1] In the 1940s and 1950s financial instability was endemic, inflation high and growth was linked to trends in the agriculture sector. Within this phase there were three subphases: relatively high growth during World War II stimulated by external demand for agricultural products; stagnation immediately after the cessation of hostilities in 1945 and for several years thereafter with the drop in external demand; and a resumption of growth in the second half of the 1950s.

1.2　　　　The 1960s, the second growth phase, were years of financial and political stability with an important shift in the source of growth from agriculture to public works and commerce. By the early sixties inflation was virtually eliminated. In the first half of the decade, agriculture continued to be the main stimulus of economic growth. However, by the late 1960s, agricultural growth slowed, as the sector underwent a transition, while the agricultural frontier expanded rapidly. Public works and informal trade with neighboring countries, especially Brazil, became the key contributors to growth. Improvements in the road network were substantial; radical changes took place in the energy sector. The 1960s also witnessed the construction of the country's first hydroelectric plant and the signing of agreements that led to the construction of the largest such plant in the world, Itaipu.

1.3　　　　In the 1970s, the third phase, growth accelerated tremendously with the construction of Itaipu. The works, which cost more than four times Paraguay's annual GDP, were carried out mainly between 1973-81 and were financed externally through the Itaipu Binational Entity (with debt guaranteed by Brazil). Non-Itaipu investment also increased. However, inflation began to accelerate, even though the balance of payments continued to show a healthy surplus. As in the sixties, the construction and commerce sectors showed the highest growth rates. Agriculture's performance also was exceptional, reflecting the expansion of the frontier begun in the 1960s.

1.4　　　　Unfortunately, investments carried out with the transitory resources flowing into Paraguay during Itaipu's construction were not prudently made. Thus there was no real base for offsetting the post-Itaipu letdown. The 1980s were years of macroeconomic instability and economic stagnation. Only at the end of the decade did the economy begin to revive with a recovery in agriculture and substantial improvements in macroeconomic policy management. Despite the slowdown of the 1980s, Itaipu dramatically improved Paraguay's growth potential and will continue to provide great benefits--among them, a substantial inflow of foreign exchange--for many years to come.

Instability and "Stagflation" in the Forties and Fifties

1.5　　　　Between 1938 and 1946, GDP growth averaged 2.5 percent per year, led by export growth of 8.2 percent per year (See Table 1.1). During World War II, demand for Paraguayan agriculture products increased sharply. During this period, the volume of tobacco exports tripled,

---

[1] A word of warning is in order. Information on Paraguay, specially trade and output related data, often are unreliable due to a large but not well-quantified, informal sector that leaves few traces on official data. The porosity of the frontiers means that large shifts to and from registered trade may occur. With unregistered trade only imperfectly estimated, this means that both trade and output figures may show year-to-year variations that are much different than the underlying reality.

wood multiplied five times, meat exports doubled, cotton and hides increased 50 percent, and vegetable oils increased more than six times (Ugarte, 1983, Annex 1, p. 305). However, exports of yerba mate showed only a slight increase, and exports of quebracho--a product taken from a tree growing in the Chaco and used to tan leather--showed only modest growth after World War II due to competition from a less expensive alternative from Africa. Inflation accelerated in this period; a steady rise in prices from 1939 to 1941 was followed by steep increases until 1944. Between 1939 and 1944 the cost of living skyrocketed 300 percent for higher income groups and 50 percent for the poor (Warren 1949, p. 334). In November 1943, the *guarani* became the country's monetary unit, at an initial exchange value of G3.07 to the dollar (one guarani replaced 100 old pesos) (Ibid., p. 335).

Table 1.1: PARAGUAY - GDP BY EXPENDITURES 1938-1988
(Growth Rate in Percent*)

	1938-46	1946-52	1952-60	1960-65	1965-70	1970-81	1981-88	1989-90
Government Consumption	2.8	4.8	6.7	1.2	8.5	6.5	2.3	4.3
Private Consumption	2.5	2.5	3.9	2.1	5.1	7.6	2.7	1.3
Gross Fixed Investment	4.4	1.5	13.8	7.0	7.5	18.2	-2.5	10.4
Exports	8.2	7.2	-7.1	9.6	3.4	9.9	9.6	16.9
Imports	9.1	11.3	-2.3	-5.5	9.1	13.5	5.3	9.0
GDP	2.5	2.4	3.8	4.6	4.8	8.9	1.9	4.4
Memo Items								
ICOR	0.80	13.77	1.40	1.60	2.57	2.14	10.81	4.6
Fixed Inv/GDP	2.8	3.9	4.8	7.7	12.0	18.3	21.8	21.2

Source: Statistical Annex, Table 2.6.
*Growth rates estimated by exponential regression.

1.6 The end of World War II brought a drastic drop in demand for agricultural products. GDP dropped 13 percent in 1947, a consequence of the end of the war and the revolution that began earlier that year. The problems caused by low export demand and social unrest that disrupted production continued into the early 1950s. They were further exacerbated by imprudent financial policies. In 1952, real GDP was about the same as in 1946, with most exports at lower levels. Not surprisingly, investment was low in this period. In an attempt to encourage production, credit policies became too expansive, fueling inflation and draining foreign reserves (Birch, n.d., p.15). Inflation accelerated from 20 percent in 1947 to more than 50 percent in 1951, and then jumped to 160 percent the following year (IBRD 1954, Statistical Annex, Table IV). A system of multiple exchange rates and exchange controls gradually developed. In 1955, the different exchange rates varied between G21 and G65 to the dollar. By the end of 1955, Paraguay had exhausted its foreign reserves (Baer and Birch 1987, p. 603).

1.7 Growth recovered somewhat in the latter half of the fifties, averaging 4 percent per year between 1952 and 1960. Commerce and construction were among the most dynamic activities (see Table 1.2). In agriculture, traditional products such as cotton, corn, tobacco, and forestry did not improve much; the bulk of the change in agriculture output came from sugarcane and manioc (WB 1965, Tables 4a, 7a). The buoyant construction sector, which grew at an average rate more than twice that of the economy, owed its dynamism to a tripling of the road network between 1954 and 1963. Investment increased faster than GDP (see Table 1.1).

However, the realized ICOR (ignoring depreciation) was a fairly high 1.4, despite a rather low growth rate and a long payoff period for the projects undertaken. Inflation declined in the second half of the 1950s, but at close to 18 percent per year in 1960, it was still high. The problem was brought fully under control after 1962 when inflation dropped to less than 5 percent.

Table 1.2: PARAGUAY - GDP BY SECTORS
(real growth rates in percent)

	1952-60	1960-65	1965-70	1970-81	1981-88	1989-90
Agriculture	2.4	4.0	1.8	6.9	2.4	4.9
Industry	1.8	6.6	6.5	8.4	1.6	4.2
Construction	9.3	6.6	6.5	21.1	-1.3	8.0
Services	6.1	4.2	6.3	9.3	2.2	4.7
Basic	2.4	3.3	4.9	11.6	4.7	6.4
Other	6.5	4.6	4.8	8.9	1.9	4.4
GDP	4.0	4.6	4.8	8.9	1.9	4.4

Source: Statistical Annex, Table 2.2
* Average growth rates estimated by exponential regression.

1.8   Important agreements that paved the way for growth recovery in the second half of the fifties--and for faster growth in the future--began in this period. Several treaties oriented to improve transportation were signed with Brazil (see the Transport section in the next chapter). In January 1956, Paraguay signed an agreement with Brazil, whereby the latter offered to finance and conduct studies of the hydroelectric potential of the Acaray and Monday Rivers, both tributaries to the Alto Parana River, close to the area where the transport projects were being advanced. Brazil also offered to cosign whatever loans were necessary for the construction of the hydroelectric plant and to buy 20 percent of the energy produced (Birch n.d., p.19). Relations with Argentina also improved. In 1958, Argentina and Paraguay signed an accord to study the hydroelectric potential of the Yacireta-Apipe rapids on the lower Parana Rivers. The report was submitted in 1964 but further studies were considered necessary (Ibid., p. 21-22).

The Sixties: Establishing the Basis for Faster Growth

1.9   Growth accelerated to 4.7 percent per year in the 1960s; however, for the decade as a whole, agriculture was not this time the main source of growth. In the first half of the decade, galvanized by rapidly increasing exports, agriculture expanded as fast as GDP (see Table 1.2), only to stagnate when export markets declined.[2] In the second half of the sixties, public sector works seem to have remained a main source of growth; construction increased at annual rates well above GDP's, while electricity and water growth began to exceed GDP's. Commerce was another important growth source, expanding by more than 6 percent per year, pushed up by regional tourism, the most dynamic element in Paraguayan exports in those years. Tourists from

---

[2] No explanation has been forthcoming for the slow agricultural growth in these latter years. The dynamic elements that fueled growth in the 1970s were to some extent in place in the second half of the sixties, but the economy did not respond to them at the time. Again caution is necessary in drawing firm conclusions because of the poor data quality.

Argentina, and especially Brazil, were attracted by improved roads and the availability of luxury goods at relatively low prices (WB 1971, p. 31).[3]

1.10   Driven by public capital outlays, fixed investment grew faster than GDP, its share frequently over 12 percent in the second half of the decade. With an ICOR around 2.6, the implicit productivity of capital seems to have declined but the ICOR remained much lower (productivity appears to have been higher) than in other countries in the early stages of development. Public investment averaged 5.3 percent of GDP in 1966-70, double the rate of the previous five years. The Government had started to carry out important programs, especially road construction, hydropower development, expansion of port facilities, improvement of water services in Asuncion, and construction of a cement plant (WB 1971, p. 39). Foreign financing for these expenditures was important--even more so was the increase in public savings. Taxes went up from 7.5 percent of GDP in 1964 to close to 11 percent in 1969/70 (WB 1971, Table 5.1). In part this increase merely reflected a shift in oil revenues to the category of taxes, once the oil refinery was installed in 1966, but it also reflected the introduction of a sales tax in 1969, higher rates for the stamp tax approved in 1967, and a further increase in 1969.[4]

1.11   Among public investments, aside from transport projects, energy ventures to take advantage of the country's hydroelectric potential are the most striking. During the 1960-70 period, the Electricity Company (ANDE) increased its generating capacity from 22 to 126 MW. Construction of the Acaray plant was carried out in the second half of the 1960s; in 1969, ANDE installed the second 45 MW generator. The second stage of this project, Acaray II, was completed in the 1970s, raising the project's capacity to 194 MW (Ugarte 1983, p. 267). This endeavor, however, was overshadowed by the signing of agreements for construction of Itaipu. Studies for this latter project began in 1961, when Brazilian President Janio Quadros commissioned the first full study of the Guaira Falls (completed in 1963). This study proposed a plant with a capacity of 10,000 MW. Construction was initially delayed because of border disputes between Brazil and Paraguay.[5] However, a solution was soon reached; in June 1966, Paraguay and Brazil signed the Acta Final, which became the basis for the Itaipu Treaty signed in April 1973 (Birch, n.d., p. 24). The Acta stated that the electricity generated would be divided equally between the two countries.

1.12   Expanding the agricultural frontier was the Government's top priority, pursued with great success during the 1970s. Although attempts to expand this frontier began soon after the Triple Alliance War (1870), by the 1950s, more than 60 percent of the country's population still was concentrated within 100 miles of Asuncion (Baer and Birch 1984). Minifundia remained

---

[3]   The catalytic element was the opening of the bridge connecting Ciudad del Este in Paraguay and Foz d'Iguazu in Brazil (para. 2.30).

[4]   It is instructive to recall that after one year of deliberations, Congress passed an income tax reform approving a global personal income tax that failed to be applied. Also, the jump in the tax revenues in 1969/70 did not last. As early as 1973, tax revenues fell below 8.5 percent of GDP. In other words, as in many other countries and circumstances, higher tax rates had only a temporary effect on collections.

[5]   In June 1965, Brazil moved troops to the area and Paraguay protested (Birch, n.d., p.23).

the typical form of land exploitation within that perimeter. For example, in 1962, Cordillera, one of the minifundia departments, had a population of 38 persons per km^2 while the national average was 4.5. With the land opened by the new roads and the founding of the IBR (Instituto de Bienestar Rural) in March 1963, 42,000 families were resettled between 1963 and 1973 (Baer and Birch 1984). However, agriculture output did not respond immediately to the new settlements--instead, it stagnated in the second half of the 1960s for reasons that are not altogether clear.

Itaipu: The Basis for Accelerated Growth in the 1970s

1.13   Between 1970 and 1981 GDP growth averaged a strong 8.9 percent per year. Agriculture grew rapidly (6.9 percent per year), as did most other sectors. Basic services like electricity, water and sewage, and transport averaged 11.6 percent per year; however, the thrust of the added dynamism was in another sector and transitory--construction expanded at an average rate of 21 percent per year. Obviously this growth could not be sustained, a fact that became increasingly clear throughout the 1980s. The dramatic increase in construction stemmed from a large expansion in private investment (public investment remained at about 5 percent of GDP). Private investment, which represented less than 10 percent of GDP (at 1982 prices) in the 1960s, increased to almost 17 percent on average during the 1970s and over 20 percent between 1979 and 1981. The ex-post ICOR fell to 2.1, despite the increase in investment, because of the extremely rapid growth rate. Public finances remained strong in this period and huge increases occurred in foreign reserves, which grew from less than US$20 million at the beginning of the seventies to US$800 million in 1981.

1.14   Inflation accelerated in the 1970s, following worldwide trends. Domestic and international (US) inflation moved together between 1970 and 1978 (see Table 1.3), reflecting a fixed exchange rate. Massive inflows of foreign exchange occurred as domestic credit policy was <u>relatively</u> tight until the end of the decade.[6] Domestic inflation slightly exceeded the international rate in 1973 and 1974. In 1974, a parallel and free market for foreign exchange developed, with a small premium on the free over the official rate. The premium almost disappeared in 1977, the free rate dropping to the official rate. However, domestic inflation substantially exceeded the international rate for the three consecutive years, 1979, 1980, and 1981, a break from the close linkage of the 1970s. A large premium developed in the free market after 1980 and in 1984, fueled by excessive credit expansion,[7] and, in the face of the diversion between domestic and international inflation rates, the official exchange rate could not be maintained (Baer and Birch 1987, p. 610; Baer and Breuer 1987, p. 134). The Government chose to initiate a complex and distortionary system of multiple exchange rates, which was eliminated only with the adoption of a unified free exchange rate in 1989.

---

[6] Between 1970-78 credit increased at an annual rate of 13 percent, well below nominal GDP growth rate of 18 percent per annum.

[7] Between 1978 and 1984 credit increased at an average annual rate of 26.5 percent, higher than the 22.1 percent average increase in GDP.

Table 1.3: PARAGUAY - DOMESTIC & INTERNATIONAL INFLATION IN THE SEVENTIES
(percent per year)

Year	International a/	Domestic b/
1970	6.3	1.6
1971	5.4	5.7
1972	9.0	8.5
1973	15.8	20.4
1974	21.8	23.5
1975	11.2	5.7
1976	1.4	4.8
1977	9.8	11.0
1978	15.1	9.9
1979	13.3	19.9
1980	9.7	16.8
1981	0.4	16.3

a/ Source: World Bank, MUV index.
b/ Source: GDP Deflator, BCP

1.15   The two most important factors in explaining the changes in growth that occurred in the seventies were the expansion of the agriculture frontier and the construction of Itaipu.

1.16   The Stroessner Administration's efforts to expand the agriculture frontier in the mid-1960s and 1970s paid off handsomely in the 1970s. The settlements of the 1960s were the main reason for agricultural growth. By the end of 1976 almost 90,000 land titles had been issued, covering about 4 million ha (WB 1978, p. 8). In response to powerful economic incentives, a large number of Brazilian farmers came to the expanding frontier to join Paraguayan colonists in the seventies as well. The frontier lands were mainly used to produce export crops, especially cotton and soybeans, which came to dominate the country's exports. The areas allocated to cotton increased five times between 1970 and 1980 (from 50,000 to 250,000 ha), and land under soybean cultivation grew from 40,000 to 430,000 ha. Other crops also benefitted from increased availability of land; areas under maize and manioc increased significantly (from 190,000 ha in 1970 to 380,000 ha in 1980 for the former, and from 100,000 ha to 140,000 for the latter) (WB 1984, Table 3). Output grew correspondingly. In that period, output of cotton, soybeans, maize, and manioc increased 535, 1,740, 230, and 10 percent, respectively (Ugarte 1983, p. 280; WB 1981, Table 7.3; WB 1984, Table 8.2). The structure of exports reflected the dramatic change in production. Cotton and soybeans represented only 6 percent of exports in 1970 but 60 percent in 1981 (Baer and Birch 1984, p. 787).

1.17   The highways to Brazil, opened in the sixties and seventies (see section on transport), also were a stimulus in agricultural growth. Paraguay's traditional dependence on Argentina as its primary trade route to the sea was broken. This was reflected by the dramatic increase in trade with Brazil, to the detriment of trade with other countries, especially the United States. Exports to Brazil were 1.6 percent of the total in 1970 and 18.3 percent in 1981 (33 percent in 1989); for imports these values were 3.2 in 1970 and 25.9 percent in 1981 (close to 30 percent in 1989). The corresponding totals for the United States dropped from 27 percent to 5 - percent for exports and from 24 percent to 10 percent for imports.

1.18   The radical changes brought about by opening new frontiers can also be observed in the regional population distribution. While only 18.3 percent of the population lived in the eastern frontier region in 1962, 27.3 percent lived there 20 years later. Also, about 40.5 percent of the population lived in the minifundia region in 1962, compared to 34.2 percent in 1982 (see Table 1.4). Contrary to migration elsewhere, the main migratory movements in Paraguay did not involve flows from rural to urban areas but rather movement out of rural minifundia into newly opened rural areas (Baer and Birch 1984, p. 788).

Table 1.4: PARAGUAY - REGIONAL DISTRIBUTION OF POPULATION
(Percent)

Region	1962	1972	1982
Asuncion	15.9	16.5	15.1
Minifundia Region a/	40.5	35.7	34.2
Eastern Frontier Region b/	18.3	22.2	27.3
Other	25.3	25.6	23.4
Total	100.0	100.0	100.0

Source: Baer and Birch 1984, p. 788, Table 5.
a/ Includes Departments of Cordillera, Guaira, Paraguari and Central.
b/ Includes Departments of Caaguazu, Itapua, Alto Parana and Canendiyu.

1.19   The other major source of growth in the seventies was the construction of Itaipu, a massive hydroelectric dam, managed by a binational enterprise with an equal number of Paraguayan and Brazilian directors. The dam was planned and actually built based on loans contracted by the binational entity in international markets, guaranteed by the Government of Brazil. The electricity generated is sold to electric companies in both countries at prices set in US currency (Baer and Birch 1984, p. 789).

1.20   Itaipu's cost was US$18.5 billion (measured at the end of 1989), four times Paraguay's 1989 GDP. Of Itaipu's cost, US$250 million (equivalent to 6 percent of GDP) was spent in Paraguay each year between 1977 and 1980 (Baer and Birch 1984, p. 789).[8] This inflow increased liquidity and helped stimulate a tremendous credit expansion while putting downward pressure on the real exchange rate. Construction benefitted most from the easier credit policies, growing at an average annual rate of 23 percent between 1973 and 1981. This reflected not so much the construction of Itaipu as the boom in private construction and public spending on infrastructure. Easy credit also helped expand commercial activities. For example, in 1972 there were only six banks in Asuncion; by 1981 there were 20. Also, plentiful foreign exchange allowed abundant import of consumer goods, and many well-known brands could be found in Paraguay. As a result, commerce and finance grew at a 10 percent average annual rate between 1973 and 1981. It is likely that agricultural growth during this period was a byproduct of increased demand stimulated by the construction of Itaipu--with the expansion of the frontier as another consequence of Itaipu, and not an independent cause of agricultural growth.

1.21	Many private sector investments[9] during this period later proved to be overdimensioned; construction investment certainly was greatly exaggerated. Initially, the public sector behaved conservatively and increased its savings; the Central Bank accumulated close to US$800 million in foreign reserves. Unfortunately, this public sector approach did not last long, as discussed below. As in a typical case of Dutch Disease, neither the private nor public sector spent the transitory income increase wisely.

The 1980s - Macroeconomic Instability and Stagnation

1.22	After 1981, the large inflow of foreign exchange from Itaipu essentially disappeared. This worsened the balance of payments and substantially reduced income for many Paraguayans, causing drops in output, employment and aggregate demand. As noted above, private investments made during the 1970s proved to be unproductive, but the public sector had accumulated considerable savings in highly liquid foreign reserves. This encouraged the Government to intervene in the economy during the 1981 downturn. The Government increased expenditures and entered (or expanded its presence) into sectors such as cement and steel. To some extent, these expenditures were financed abroad; but expansive credit policies also were used.

1.23	The outcome of these actions differed substantially from Government expectations: investments turned out to be overdimensioned for the domestic market and targeted to sectors where regional markets already showed substantial excess capacity; hence, most of this investment was, and remains, unproductive. Foreign debt increased dramatically, repayment conditions were not met, and after a time the Government began to accumulate substantial arrears. Even the increase in foreign debt and external arrears was not enough to finance public expenditures; monetary expansion rose and inflation became rampant. Between 1982 and early

---

[8]	For national accounts purposes, Paraguay does not consider binational enterprises as being located within the national territory; therefore, their accounts (investment, value-added, etc.) are not part of GDP. Sales of Paraguay to them are included as exports (and in that sense part of GDP); their imports are not included in the national accounts; and wages paid to Paraguayans and royalties and other Paraguayan income are considered factor payments received by Paraguay, part of GNP not GDP.

[9]	Investment figures exclude all direct expenditures linked to Itaipu as discussed in note 8.

1989, the Monetary Authorities lost a large percentage of the reserves accumulated during the construction of Itaipu. In an attempt to counteract the balance of payments difficulties, the Authorities imposed a system of multiple exchange rates, under which, ultimately, the public sector became the main legal beneficiary. These multiple exchange rates not only did not resolve the balance of payments difficulties; they also distorted incentives. And with the higher inflation, distortions caused by the multiple exchange rates increased. This, in turn, affected growth and started the vicious circle of policy instability, inflation, and stagnation, which Paraguay had been able to avoid since the late fifties. Thus, Government policies not only failed to ameliorate the slowdown but induced instability that exacerbated the effects of the recession. Some progress was achieved between 1986 and 1989, but it was too little, too late.

1.24   **Growth by Sector in the 1980s.** Not just growth but absolute GDP declined for two consecutive years after the completion of Itaipu's civil works in 1981. The economy's absolute contraction (GDP was 4 percent lower in 1983 than in 1981) stopped in 1984; however, this was not enough to avoid additional output loses in 1984-86. Growth remained at less than 2.5 percent per year for the next three years, well below the 3 percent p.a. population increase. A recovery began in 1987 and accelerated in 1988, when growth exceeded 6 percent. However, this still did not bring income back to levels that would correspond to a long-run trend of about 5 percent growth per year. According to national accounts figures, exports continued to expand fairly rapidly in the 1980s, though not nearly as fast as in the 1970s (see Table 1.5). Domestic demand stagnated, reflecting the slowdown in construction and services and the loss of factor income from Itaipu, and so did GDP.

TABLE 1.5: PARAGUAY - GROSS DOMESTIC PRODUCT
(real growth rate in percent)

	1971-81a/	1981-86a/	1987	1988	1989	1990
Total Expenditures	9.7	0.2	3.0	6.9	4.3	2.1
Consumption	7.7	2.2	2.2	7.7	2.9	0.2
Public	7.2	0.7	10.0	3.3	5.4	3.2
Private	7.7	2.3	1.6	8.1	2.6	-0.1
Investment	18.7	-6.5	6.3	3.7	10.7	10.1
Imports	13.9	0.6	19.4	8.7	14.0	4.3
Expend. in Dom. Goods	8.9	0.1	-0.8	5.4	1.6	1.4
Exports	10.8	5.8	36.7	6.0	25.1	9.3
GDP	9.2	0.8	4.3	6.4	5.8	3.1

Source: Statistical Annex, Table 2.6.

a/ Average growth rates estimated by exponential regression.

1.25   The sectors benefitting most from the Itaipu boom--construction, and commerce and finance--were those that suffered most in the post-Itaipu period (see Table 1.6). Construction activity in 1989 was still more than 7 percent below that of 1981; although trade and finance was 16 percent higher in 1989 than in 1981, this sector's growth rate was below GDP's for all years after 1981 except two (1985 and 1986).

TABLE 1.6: PARAGUAY - GROSS DOMESTIC PRODUCT
(Real growth rates by sectors in %)

	1977-81 a/	1981-86 a/	1987	1988	1989	1990
Agriculture	7.5	1.1	7.0	12.1	7.7	2.2
Industry	10.4	0.5	3.6	5.9	5.9	2.5
Construction	26.4	-2.9	2.0	2.6	2.5	-0.9
Services	11.2	1.3	3.5	4.1	5.1	4.2
Basic b/	11.3	3.9	6.0	7.1	5.3	7.6
Other	11.2	0.9	3.1	3.5	5.1	3.6
GDP	10.8	0.8	4.3	6.4	5.8	3.1

Source: Statistical Annex, Table 2.2.

a/ Average growth rates estimated by exponential regression.
b/ Basic services include transport, electricity and water.

1.26   Agricultural stagnation in the eighties, as reflected in the national accounts, involved a significant restructuring of output oriented to the domestic market: an impressive expansion in corn and wheat--corn doubled and wheat increased five times between 1981 and 1987--almost compensated for reductions in traditional products like manioc, tobacco, coffee, and bananas. Agriculture's poor performance contributed to commerce's slow growth. Banking services also stagnated, reflecting the substantial reduction in financial savings.

1.27   Manufacturing output moved at about the same pace as global GDP. Agriculture-related industries, mostly oriented to the domestic market, continue to comprise most of the country's manufacturing activities, with food, beverages, tobacco, cotton (including textiles), woods, and hides account for more than two-thirds of the sector. As might be expected, in a sector where exports are insignificant, manufacturing stagnates along with domestic demand. The steel and cement industries' performance also reflects slow domestic growth and regional excess capacity. As a result there is substantial excess capacity in these sectors.

1.28   As in earlier decades, basic services--electricity, water and sewage, and transport--expanded faster than the economy; but growth also declined substantially compared to earlier years. Transport services expanded the least in the general economic downturn. Electricity, water and sewage continued to grow at rates over 6 percent a year through the eighties. In particular, the beginning of operations at Itaipu in 1985 promoted high electricity growth rates after 1985; the completion of important water and sewage works in Asuncion explains the higher growth rates for this sector until 1983.

1.29   It should be noted, however, that data problems make developments in this period difficult to interpret. For example, between 1981 and 1986, national accounts indicate that the real value of total exports of goods advanced at a relatively fast rate (5.8 percent per year), while exports of goods registered by the Central Bank (or exports measured by imports of other countries, which is the basis of the balance of payments estimates[10]) declined. Thus the increase

---

[10] However, balance of payments estimates may be more unreliable than the national accounts. For a more complete explanation see paras. 1.45-1.47 below

in the total is wholly due to the estimated evolution of unregistered exports, about which little is known with any certainty. Although exports might not have increased as much as national accounts estimates suggest, this report's hypothesis is that such evolution was indeed plausible.

1.30   National accounts figures for specific agricultural products suggests a modest if not negative real growth of export of goods between 1981 and 1986, which probably understates what really happened. Cotton and soybeans have been the economy's most dynamic products in the last two decades, and because most of their output is exported, little change might be expected in the share of output exported. However, the drastic changes in this share for both cotton and soybeans, as depicted in Table 1.7, suggest that registered exports and corresponding output figures are incompatible. Significant changes in incentives to export through informal channels make registered exports a poor proxi for total exports. Moreover Paraguay's exports may vary from year-to-year because of reexports from Brazil, which may or may not be registered. Exchange rate policy discouraged non-registered exports before 1981, and encouraged them afterwards;[11] only in 1989 conditions were again like in 1981. In 1989, cotton and soybeans represented nearly 70 percent of registered exports and 37 percent of agriculture exports; the corresponding figures for 1981 were only a little lower, 60 percent of registered exports and 31 percent of agriculture output. But yearly changes within the 1981-86 period show a different picture. For example, between 1981 and 1982 cotton's domestic and external demand increased strongly--the volume of cotton (registered) exports increased 23 percent and textile output increased 20 percent--but domestic output is estimated to have declined 25 percent. The observed high ratio of output exported shown in Table 1.7 may be misleading; output growth maybe understated (or unregistered exports from Brazil may have increased). Between 1982 and 1986 the contrary happened; (registered) exports were constant and textile output increased barely 7 percent while estimated cotton output jumped 23 percent; thus the share of output exported declined, another misleading change. This time, cotton exports' growth probably was underestimated by the evolution of registered exports. Registered exports as a whole declined 26 percent between those years but non-registered exports increased strongly, so that the total of registered and unregistered went up by more than 50 percent. Because of the exchange rate policy, the evolution of both total and non-registered exports after 1982 seems plausible. Therefore, the trend of overall exports of goods is inaccurately estimated by registered exports. If estimates of overall exports are right and the evolution of non-registered exports of cotton explains the evolution of the aggregate, it could well be that cotton output might have grown as much as stated between 1982 and 1986; still, because of the 1982 bias, output growth probably is underestimated in the 1981 and 1986 period. The picture offered by soybeans output is more clear-cut and suggests inaccuracies like cotton's. Soybeans are almost fully exported (although national accounts indicate that only 25 percent of the country's output was sold abroad in 1981-- see Table 1.7). Between 1981 and 1986 exports' volume increased almost 200 percent, but domestic output is estimated to have declined more than 20 percent. Even though the fraction of output sold abroad appeared at a more reasonable 90 percent in 1986, output growth seems to have been substantially underestimated between 1981 and 1986. Thus in both of these important products, cotton and soybeans, output growth between 1981 and 1986 might have been underestimated in national accounts. GDP growth would have been understated correspondingly.

---

[11]   For a more detailed discussion of exchange rate policies, see the balance of payments section, paras. 1.43 - 1.50.

Table 1.7: PARAGUAY - SHARE OF EXPORTS IN DOMESTIC OUTPUT

	Soybeans	Cotton
1970	0.0	30.1
1971	16.2	17.4
1972	41.5	18.8
1973	44.5	22.1
1974	53.0	20.6
1975	46.3	26.5
1976	74.4	31.0
1977	68.9	26.4
1978	58.2	29.4
1979	74.2	33.3
1980	36.2	32.1
1981	25.2	28.6
1982	62.3	44.0
1983	81.0	34.6
1984	64.5	27.9
1985	72.4	34.5
1986	90.1	35.4
1987	93.6	37.3
1988	89.9	31.2
1989	115.4	35.1

Source: Central Bank of Paraguay, National Accounts.

1.31   **Investment and Savings**. The country's overall capital expenditures dropped substantially after 1981. Investment expenditures increased at a 21 percent annual rate during Itaipu years (without considering the works themselves which appear as part of the Binational Corporation); they contracted more than 6 percent per year between 1981 and 1986. Investment which had reached 26 percent of GDP in the 1977-81 period, at the height of works in Itaipu, dropped to 22 percent in 1982-86 (see Table 1.8). Private investment in particular declined sharply in the 1980s. Public investment remained high; indeed on a cash rather than national accounts basis, it was somewhat higher in the period 1982-86, than in 1977-81 (see Table 1.9).

1.32   National savings fell even more than investment according to national accounts. The share of national savings in GDP dropped from almost 22 percent to less than 13 percent between 1977/81 and 1988. This was the result of low internal interest rates, not only negative in real terms (they had also been negative in prior years) but extremely uncompetitive with rates abroad. With the completion of most civil works in Itaipu, foreign savings (i.e., the current account deficit) increased substantially after 1980/81.[12]

1.33   This picture changed when the new Administration took office in February 1989. The policy changes undertaken produced some recovery in private investment and a strong increase in national savings both from public and private sources. National savings grew to 17 percent of GDP. In 1989-90 there was a radical cut both in the private and public deficits and a corresponding reduction in the current account deficit (foreign savings), from 10 percent down to 5 percent of GDP.

Table 1.8: PARAGUAY - SAVINGS AND INVESTMENT, 1977-1990
(Percent of GDP)

	1977/81	1982/86	1987	1988	1989	1990
Gross Fixed Invest.	26.1	22.0	23.7	23.1	22.7	22.0
Private	19.7	16.6	16.6	16.6	19.1	18.5
Public	6.4	5.4	7.1	6.6	3.6	3.5
Gross Savings	26.1	22.0	23.7	23.1	22.7	22.0
National	21.7	14.8	14.0	12.9	16.2	17.0
Private	15.4	13.5	12.5	11.7	11.2	9.1
Adjusted Public	6.3	1.3	1.5	1.2	5.0	7.9
Public	5.1	-0.1	-0.2	-0.3	3.0	5.6
Inflation Tax	1.2	1.4	1.7	1.5	2.0	2.2
Foreign	4.4	7.1	9.7	10.3	6.5	5.0

Source: Statistical Annex, Table 2.6.

1.34   **Public Finance Policies**. As noted above, in an attempt to counteract the depressive effects of lower expenditures at Itaipu, the Government increased its investment, which went up from an average of 5.6 percent of GDP between 1980 and 1982 to 6.8 percent between 1983 and 1988 (7.3 percent excluding 1986--see Table 1.9); the increase mainly reflected the construction of a public cement plant and a steel mill. In addition, public finances suffered as tax collection deteriorated as a result of a declining economy and poor tax administration. After averaging 8.1 percent of GDP in the 1980-82 period, revenues dropped to 6.8 percent between

---

[12]   Foreign savings in national accounts are not consistent with balance of payments estimates. In the former, the movements in foreign and national savings are more consistent with related policy changes. The balance of payments estimates suggest that foreign savings would have dropped strongly from 1981 to 1983 and increased dramatically from 1983 to 1987, which is not consistent with policies in place; the errors and omissions account also shows large and counterintuitive changes in the opposite direction. In fact, if the errors and omissions account is aggregated with the current account, foreign savings estimated from the balance of payments become more plausible and get closer to the national accounts estimates. Nonetheless, substantial discrepancies remain (see paras. 1.46 and 1.47).

1983 and 1988. The Government tried to counteract these negative effects on public finances by cutting "formal" current expenditures, excluding interest payments, which dropped from 11 percent of GDP in 1982/83 to 9 percent after 1984. However, those cuts had little impact, since they were offset by a rapid rise in interest payments (from 0.5 percent of GDP in 1980/81, to 1.5 percent in 1987/88), even with the accumulation of external arrears.

1.35   Although formal public sector official expenditures (excluding the financial sector) were cut, effective outlays probably increased because official expenditures in foreign currency were valued at an artificially low exchange rate. In fact, a major reason for the deterioration of public finances in the latter half of the 1980s was the larger (effective) expenditures that were financed by foreign exchange subsidies. The subsidies cannot be disaggregated into current or capital expenditures for analytical purposes. However, they are included in this report's estimate of the deficit to be financed (See Table 1.9 including note b).

Table 1.9: PARAGUAY - CONSOLIDATED PUBLIC SECTOR ACCOUNTS (CASH BASIS a/)
(% of GDP)

	1980	1981	1982	1983	1984	1985	1986	1987	1988	1989	1990p/
Current Revenues	14.6	13.4	15.1	14.3	13.1	14.1	13.7	14.6	14.3	17.3	18.9
Taxes	8.5	7.6	8.3	6.5	6.7	6.9	6.8	7.1	7.0	8.8	9.2
Value Added Ent. b/	2.9	2.6	3.1	3.4	2.5	3.5	3.1	3.8	3.4	3.1	4.3
Other	3.2	3.2	3.7	4.3	3.9	3.7	3.7	3.6	3.8	5.4	5.4
Current Expend.	9.3	10.2	12.3	12.7	11.1	10.2	10.3	10.8	10.0	13.8	13.3
Non Interest	8.8	9.7	11.3	11.8	10.1	9.3	9.4	9.3	8.6	10.6	11.2
Interest a/	0.5	0.5	1.0	0.9	1.0	0.9	0.9	1.5	1.4	3.2	2.1
Current Savings	5.3	3.2	2.8	1.6	2.0	3.9	3.3	3.8	4.3	3.5	5.6
Capital Expend.	5.3	6.1	5.5	6.9	8.4	6.5	5.1	6.7	7.4	6.8	5.0
Official Deficit a/	-0.1	2.9	2.7	5.3	6.4	2.6	1.7	2.9	3.1	3.3	-0.7
Forex Subsidies c/	0.0	0.0	0.8	0.6	2.7	4.9	5.5	4.0	4.5	0.5	0.0
Deficit = Financing	-0.1	2.9	3.4	5.9	9.1	7.5	7.2	6.9	7.6	3.8	-0.7
External Financ.	0.8	0.9	1.0	3.5	4.2	0.9	2.7	0.5	0.6	4.3	-2.0
Domestic Fin.	-0.8	2.0	2.4	2.4	4.9	6.6	4.6	6.4	7.1	-0.5	1.3
CB Forex Sub	0.0	0.0	0.1	0.1	1.8	2.4	3.7	3.7	3.0	0.3	0.0
Banking Credit	0.3	1.7	0.7	2.3	0.3	0.8	0.4	-0.3	-0.8	-2.4	-0.5
Non Bank Forex Sub	0.0	0.0	0.7	0.5	0.9	2.5	1.8	0.3	1.5	0.2	0.0
Other	-1.2	0.4	0.9	-0.5	1.9	0.9	-1.4	2.7	3.4	2.4	1.8

Source: Finance Ministry and Bank estimates.

a/ Arrears not included.
b/ Value of sales less cost of inputs. Public enterprises' other own revenues are included in "other"; thus the "volume added" concept in this table is not the same as in section 3C.
c/ The full counterpart of these subsidies is included in the deficit to be financed. To some extent this is an overstatement; the fraction financed by the Central Bank is money creation, but that financed by exporters (and importers of capital through official channels) represents a transfer of resources to importers and thus, effectively, is a tax and subsidy.
p/ Preliminary estimates.

1.36   The sizable subsidies the public sector received on its purchases of foreign exchange came both from exporters (who had to sell to the Central Bank a fraction of their export proceeds at an exchange rate well below the market rate--the "aforos") and the Central Bank (which sold foreign exchange to the public sector at even lower prices). These purchases reached the equivalent of nearly 5 percent of GDP between 1985 and 1988. Thus, although the official deficit did not go over 3 percent of GDP during the entire decade (except in 1983/84), the effective deficit, including the foreign exchange subsidy, averaged more than 7 percent of GDP in the 1983-88 period. After a small surplus in 1980, with the general economic decline after Itaipu, the public sector deficit increased to 3.4 percent in 1982 and to 7.4 percent of GDP between 1983 and 1988 (see Table 1.9). In sum, despite low official deficits, public finances performed poorly through most of the eighties.

1.37   Deficit financing introduced additional instability. Changes in foreign financing caused instability in the real exchange rate by affecting the availability of foreign exchange, and domestic financing generated inflation by forcing large monetary expansions. Foreign financing[13] fluctuated substantially: it represented 0.9 percent of GDP between 1980 and 1982, jumped to 3.9 percent in 1983/84, and dropped to 0.5 percent in 1987/88. Domestic financing showed a steady increase, as reflected in monetary aggregates. Although formal credit from the banking system to the public sector generally was small if not negative in the eighties (it exceeded 2 percent of GDP only in 1983), the banking system resources channeled to the public sector show large increases when foreign exchange subsidies are taken into account. Including these subsidies, banking credit to the public sector increased from 0.3 percent of GDP in 1980 to 2.4 percent in 1983, and to over 3 percent in the 1985-87 period (see Table 1.9). Starting in 1983, the deficit's domestic financing became a major source of disequilibrium.

1.38   The picture changed radically in 1989 with the new Administration. Against heavy odds, they improved public finances substantially. The foreign exchange subsidy (about 5 percent of GDP on the average between 1985 and 1988) was eliminated when the new economic program was put in place in February of 1989 and the multiple exchange rates unified. Moreover, the adjusted cash deficit was substantially cut from almost 8 percent of GDP in 1988 to an estimated surplus in 1990, and domestic financing (including banking system credit) was negative, after representing more than 6 percent of GDP between 1984 and 1988. Actions included increases in tax revenues, courageous cuts in expenditures, as well as improvements in public enterprise finances, all despite no new taxes.

1.39   Tax revenues as a percentage of GDP had declined from 9.4 percent in 1971 to 6.5 percent in 1983 and were only 7 percent in 1988. Mainly by improving tax administration, charging import taxes on the market value of imports as the tax base, and enacting a small tax on exports (that replaced the old "aforo" system), revenues increased to 8.8 percent of GDP in 1989 and 9.2 percent in 1990.

1.40   In 1989, cuts in expenditures and strongly enforced rationalization policies offset part of the strong upward pressures stemming from the elimination of foreign exchange subsidies (mostly spent on fuels and servicing interest on foreign debt) and higher wages and held the increase in current expenditures to two percentage points of GDP. Capital expenditures declined relative to GDP, in consonance with a general tightening of public finances and restructured priorities, which were reflected in cuts in works considered unnecessary (the excessively elaborate airport originally proposed for Ciudad del Este, for example).

1.41   Performance in the rest of the public sector (especially in public enterprises) also offered encouraging aspects in 1989 and 1990 despite unfavorable circumstances. For example, when the exchange rate was unified, the fuel bill of PETROPAR (the petroleum company in charge of oil imports) did not increase as a share of GDP.[14] In 1989, wage increases were significant for public enterprises (30 percent in May), and tariff increases were well below expectations, but total enterprise savings (excluding interest payments) did not decline as much as expected.

---

[13] Without considering arrears in interest payments, which, however, were not large. Interest arrears were less than one-third of the total arrears at the end of 1989.

[14] PETROPAR's exchange rate for import of oil (US$115 million) rose from G400 per dollar to G1000 per dollar.

1.42   Public investment was reduced in several areas in 1989, in part by reducing the waste of the mid-1980s but also because of more stringent budget constraints. The construction of a cement plant and a new steel mill had been completed by 1988--both turned out to be greatly overdimensioned for the domestic and regional market and now represent a burden on the public sector. However, because of important new projects, overall PE's investment increased in 1989 and again in 1990. Public enterprise outlays, about 3.9 percent of GDP in 1988, are estimated (preliminarily) at 5.6 percent in 1990.

1.43   **The Balance of Payments in the 1980s.**[15] The effects of the completion of the main works at Itaipu dominated the external accounts for most of the 1980s, as it did the economy as a whole. First came the drop in associated external net inflows; then a lack of adjustment to the Balance of Payments deficits. The evolution of the foreign capital account aggravated the misdirected policies: too many resources were channeled to the country in the first few years after the reduction of Itaipu inflows--which encouraged maintaining the incorrect policies--then, because of lack of progress in policy management, too few were channelled in the following years--to which the government responded by accumulating arrears.

1.44   In 1980/81 net earnings (exports, wages, and sales of expropriated land) from the binationals (Itaipu and Yacireta) totaled about US$435 million a year according to the balance of payments accounts; in 1983 they were US$315 million, and less than US$90 million in 1986 (see Table 1.10). This drop was the equivalent of over 6 percent of GDP. In 1982-83, the Government offset this reduction through increases in medium- and long-term borrowing. In 1983, these additional inflows reached almost US$150 million more than in 1981, more than compensating for the Itaipu-related losses of foreign exchange in that year. Adjustment was needed to the decline in foreign exchange earnings through changes in relative prices, but this was delayed by the borrowing and loss of foreign exchange reserves.

1.45   After Itaipu, the real exchange rate needed to increase (depreciate) to reflect the lesser availability of foreign resources. Until 1981, the real exchange had declined (the domestic currency appreciated) due to the abundant supply of foreign exchange--falling by 6-7 percent per year between 1978 and 1981. In 1981-83, the real exchange rate should have increased, reflecting a lower supply of foreign exchange. However, the adjustment was avoided by foreign borrowing. With the overly expansive public expenditure policies, not even the greater foreign borrowing was enough. The result was the country lost US$100 million in foreign reserves. Overall the real exchange rate dropped (appreciated) instead of rising (depreciating).

1.46   According to balance of payments figures, (see Table 1.10)[16], exports declined in the early 1980s, reflecting the appreciation in the real exchange rate. However, according to the balance of payments statistics, imports also fell, contrary to what might be expected with a real

---

[15]   As noted above, there are significant differences between balance of payments and national accounts data.

[16]   It must be noted that this table is not fully consistent with that shown in the projections. Debt related figures (interests, disbursements and amortizations) in the former are from official Central Bank figures; for the latter, the source is the World Bank debt statistics.

exchange rate appreciation. As a consequence, the balance of payments shows an improvement in both the resource balance and the current account.[17] This movement in imports is, however, probably an aberration in the data. The balance of payments also shows a large increase in outflows under errors and omissions. This most likely reflects (net) unreported imports. If these errors and omissions are added to imports, the current account and resource balance would more closely resemble what could be expected from the appreciation of the real exchange rate.

1.47 The policy of increasing foreign borrowing and reserve loss was unsustainable. In 1983 a distortionary and corruption-prone system of multiple exchange rates was implemented to try to stop the foreign exchange drain. As part of this system, the average exchange rate was depreciated substantially in nominal and real terms. The average real exchange rate almost doubled between 1983 and 1985 (see Table 1.10)[18]. The result should have been an improvement in net exports and the current account and resource balance. However, according to the current account, exports were fairly constant (and even declined in 1985) while imports increased. As a result the balance of payments reports a swing in the current account from +2.4 percent of GDP in 1983 to -12.7 percent in 1987.[19] Again, there seems to be a problem with the balance of payments estimates, since errors and omissions show a massive improvement from -5 percent of GDP in 1983 to roughly zero in 1985 and to +9.8 percent in 1987. It is likely that the improvement in the current account and resource balance is hidden in the capital account under the label "other" but in fact reflecting exports which went unreported to avoid the implicit tax imposed by the multiple exchange rate system and the shift of formerly unregistered imports to the official exchange rate.

1.48 Despite the improvement, the current account plus errors and omissions remained negative for the period 1984-88. Moreover, net medium- and long-term financing to the public sector dried up. These funds represented about 5 percent of GDP in 1983/84, but were negative in 1987/88. The international financial community (including the World Bank) correctly reduced voluntary disbursements for macroeconomic reasons. Adjusting the economy to the negative inflows would have meant a decline in domestic well being; instead the Government chose to accumulate arrears. Even with the accumulation of arrears, international reserves declined a further US$320 million between 1984 and 1988.

---

[17] As explained earlier (see paras. 1.29 and 1.32) balance of payments and national accounts figure are not consistent. However, the latter does suggest that the current account deteriorated in 1982-83.

[18] As noted above, the system was used as a tax to benefit the public sector, not importers or private capital.

[19] This time the national account data also show a big increase in the current account deficit.

TABLE 1.10: PARAGUAY - BALANCE OF PAYMENTS SUMMARY, 1980-89
(US$ million)

	1980	1981	1982	1983	1984	1985	1986	1987	1988	1989	1990
Exports (Gds & NFS)	944.9	965.6	781.6	742.2	756.2	671.3	790.8	806.8	1159.6	1538.6	1804.5
Binationals	266.6	322.2	218.2	245.0	156.4	105.2	62.0	75.4	141.7a/	221.2a/	225.5a/
Imports (Gds & NFS)	-1193.5	-1175.5	-919.8	-667.5	-907.3	-840.1	-1108.6	-1201.6	1320.8	1297.6	1739.0
Resource Balance	-248.6	-209.9	-138.2	74.7	-151.1	-168.8	-317.8	-394.8	-161.2	241.0	65.5
Net Factor Income	94.4	136.7	113.5	55.4	30.7	-28.8	-32.8	-88.6	-84.1	-28.8	-7.2
Accrued Interest	-77.4	-78.7	-79.9	-64.1	-78.3	-106.8	-117.0	-161.5	-138.0	-112.5	-101.1
Wages (Binationals)	100.2	122.8	81.2	66.2	47.2	19.0	27.4	45.2	23.7	39.1	33.8
Others Net	76.1	98.3	117.2	59.5	71.1	76.9	57.9	31.9	30.2b/	44.6b/	60.1b/
Net Current Transfers	4.5	5.7	5.0	6.2	9.3	7.5	11.1	27.0	35.2	23.9	32.9
Current Account	-149.7	67.5	-19.7	136.3	-111.1	-190.1	-339.5	-456.4	-210.1	236.1	91.2
Public LT (vol.)c/	148.1	133.9	225.7	281.3	209.5	81.0	139.6	-69.1	-63.5	-114.9	-108.3d/
Binationals (expropr)	28.6	25.4	21.6	5.4	9.6	1.4	1.0	10.6	-45.7	-42.0e/	2.1e/
Others (Net)	98.8	61.6	-69.4	-180.7	-109.8	-13.6	55.6	156.1	-80.8	150.5f/	-9.1d/
Capital Account	275.5	220.9	177.9	106.0	109.3	49.9	154.2	24.3	-190.0	-6.4	-115.3
E & O	38.7	-108.2	-233.2	-281.8	-93.7	29.6	59.5	367.2	196.2	-92.2	111.8
Overall Deficit	-164.5	-45.2	75.0	39.5	95.5	110.6	125.8	64.9	203.9	-137.5	-87.7
Decline Reserves	-164.5	-45.2	71.5	39.5	95.5	68.6	59.3	-46.2	143.8	-136.7	-245.9
Increase in Arrears	0.0	0.0	0.0	0.0	0.0	42.0	66.5	111.1	60.1	-0.8	158.2
Memo Items (as % of GDP)											
Resource Balance	-5.6	-3.7	-2.6	1.3	-3.4	-5.3	-9.0	-10.6	-4.1	5.9	1.2
Current Account	-3.4	-1.2	-0.4	2.4	-2.5	-6.0	-9.6	-12.2	-5.3	5.7	1.7
Errors and Omissions	0.9	-1.9	-4.3	-5.0	-2.1	0.9	1.7	9.8	5.0	-2.2	2.1
Public Net LT (volunt)	3.3	2.4	4.2	5.0	4.8	2.6	3.9	-1.9	-1.6	-2.8	-2.1
Change Reserves	-3.7	-0.8	1.3	0.7	2.2	2.2	1.7	-1.2	3.6	-3.3	-4.7
Change Arrears	0.0	0.0	0.0	0.0	0.0	1.3	1.9	3.0	1.5	0.0	3.0
Real Ex. Rate Index	100.0	86.1	88.4	83.0	109.2	157.6	140.4	139.3	139.6	147.9	130.5
GDP ($ million)	4448	5625	5419	5603	4386	3161	3547	3733	3951	4115	5264

Source: Statistical Annex, Table 3.1.

a/ Includes royalties and income from sales of surplus electricity to Brazil.
b/ Includes accrued interest on Itaipu's debt with Paraguay.
c/ Amortizations on an accrual basis.
d/ Amortizations in 1990 do not include debt reduction component in prepayment of debt with Brazil (US$316 million).
e/ Includes as outflows loans to Itaipu and its arrears of amortization and interest.
f/ Includes as inflows renegotiated arrears of interests (US$54.2 million) and amortizations (US$106.7 million) on debt with Brazil.

1.49   The balance of payments situation improved radically after February 1989, when the exchange rate was unified and "freed." Nonetheless, some problems remain to be corrected if the recent achievements are to be sustained. The real exchange rate increased in 1989, and with much tighter public finance management, there was a massive accumulation of foreign exchange reserves (over US$130 million). The sum of the current account and errors and omissions improved significantly, from a slight deficit in 1988, to a surplus of 3.5 percent of GDP in 1989. However, voluntary medium- and long-term capital inflows remained negative. Nonetheless, the capital account as a whole improved, taking into account the refinancing of arrears on the debt to

Brazil. Another problem was that arrears continued to accumulate on non-Brazilian debt. The overall balance of payments result was reserve accumulation, equivalent to more than 3 percent of GDP. This accumulation exceeded the surplus in the public sector accounts, meaning that it generated (net) monetary expansion which allowed inflation to proceed and even accelerate.

1.50   International reserves increased US$246 million in 1990, even more than in 1989. However, this was in the context of a real appreciation and a deterioration in the current account surplus (with the elimination of the multiple exchange rate system the aggregation of the current account and the errors and omissions account makes less sense--the current account alone is a reasonable measure of trade in goods and services). Net voluntary disbursements (including only Paraguay's actual payments on rescheduled debt with Brazil) were again negative (2.1 percent of GDP). Arrears accumulation was about 3 percent of GDP and there was no refinancing as in 1989. The main factor in the increase in reserves was a massive increase in errors and omission. The large improvement in errors and omissions probably reflected flows of short term capital responding to the high domestic interest rates, tight domestic credit conditions and open capital account; as in 1989 this inflow fueled monetary expansion and allowed inflation to proceed at high rates.

1.51   External Debt rose substantially in the eighties as a result of the public sector's borrowing. Medium- and long-term debt outstanding and disbursed was US$784 million in 1980 and US$842 million in 1981 (15 percent of GDP); in 1983 it was US$1.3 billion, or about 23 percent of GDP, and in 1989 it was US$2.3 billion (including interest arrears) or 62 percent of GDP (see Table 1.11). After Paraguay implemented a system of multiple exchange rates, the Government ruled that official lenders should disburse their loans at artificially low rates. As this became a source of corruption, many official lenders stopped channeling money to Paraguay (the World Bank stopped disbursements against local currency expenditures in 1980).

1.52   Terms also hardened significantly as Paraguay increased its borrowing from commercial banks. In 1980, the average interest rate was slightly higher than 5 percent, increasing to close to 8 percent in 1989; scheduled amortizations in 1980 implied an average maturity of 15 years but only 9 years in 1989. Scheduled debt service, which was US$80 million in 1981, increased to over US$400 million in 1987.

1.53   In 1989, Brazil and Paraguay reached an accord on the US$427 million the latter owed (scheduled debt service on corresponding loans was US$70 million before the deal). This debt could be serviced or prepaid with Brazilian foreign debt valued at par but purchased in secondary markets at a substantial discount. In October 1990, Paraguay had paid back that debt in full, with an average 70 percent discount.

1.54   External arrears accumulation has become a major problem. With the tightening foreign exchange situation in the mid-1980s, the Government elected to stop debt service payments in commercial bank and most bilateral official debt, but continued to repay multilateral creditors. Arrears (principal plus interest) reached US$280 million in 1989 and soared to over US$440 million in 1990 concurrently with the accumulation of foreign reserves. About one-third of arrears reflect unpaid interest.

1.55   With exports increasing strongly due to the new exchange rate policy the country's potential creditworthiness is improving. However, the problem posed by external arrears remains. Solving it will require a deal that may not be easy to reach because a significant fraction of arrears involve debt whose legitimacy the Government questions (it is linked to allegedly corrupt transactions carried out by the previous administration).

Table 1.11: PARAGUAY - EXTERNAL DEBT AND DEBT SERVICE, 1975-1989
(US$ Millions)

	1975	1980	1981	1983	1985	1987	1988	1989	1990
Long Term Debt	228.0	783.9	842.0	1273.0	1638.0	2253.0	2124.0	2125.0	1755.0
Official sources	138.6	405.0	454.6	706.0	1043.0	1324.0	1256.0	1608.0	1205.0
o/w IBRD	7.7	67.7	98.5	168.4	247.9	372.3	314.9	282.0	279.0
Private sources	89.4	378.9	387.4	567.0	595.0	929.0	868.0	517.0	550.0
Short Term Debt (inc. Int. Arrears)	..	174.0	308.0	140.9	177.8	268.0	231.0	260.0	372.0
Interest Arrears (Est.)				8.0	27.0	67.0	93.0	79.0	114.0
Long Term Debt Service Payments	30.0	124.9	131.4	101.1	158.3	225.0	341.9	311.1	334.7
Amortizations	21.5	80.4	95.0	53.4	78.0	131.0	223.9	218.6a/	250.3
Interest	8.5	44.5	36.4	47.7	80.3	94.0	118.0	92.5	87.1
Interest on Short Term Debt (Est.)	..	22.0	40.0	0.7	0.3	14.9	20.1	20.0	14.0
Total Debt Service Payments	30.0	146.9	171.4	101.8	158.6	239.9	362.0	331.1	351.4
Debt Ratios									
Long Term Debt/GDP	14.8	17.6	15.0	22.7	51.8	60.4	53.8	51.6	33.3
Debt Service Ratio b/	11.7	18.8	22.	22.1 13.7	23.6	29.7	31.2	21.5	19.5
Total Debt/GNP	14.8	21.5	20.4	25.2	57.4	67.5	61.7	58.6	40.4

Source: World Bank Debt Reporting System, Central Bank of Paraguay and World Bank estimates.
a/ Does not include rescheduled future amortizations on debt with Brazil.
b/ Total Debt Service Payments as a percent of exports of goods and services.

Credit and Monetary Policies

1.56   Between 1980 and 1988, domestic credit policies largely reflected the country's macroeconomic policy instabilities. However, beginning in 1989, monetary/credit/exchange policy must adapt to face a new challenge arising from the possible destabilizing influence of short-term international inflows.

1.57   Domestic credit policy was increasingly expansive during most of the eighties. In 1981/82, the increase in credit represented 2-4 percent of GDP (See Table 1.12). In 1981, a rapid expansion of quasi-monetary savings financed both additional credit and a strong foreign reserves accumulation while narrowly defined money (M1) barely increased. In 1982, the first year after the Itaipu boom, narrow money growth was again negligible and quasi money grew more slowly; credit expansion was "financed" largely by large losses of foreign reserves. An additional inflationary element in 1982--which would remain substantial throughout the eighties-- was a reduction in the real value of quasi-monetary savings.

1.58 Between 1983 and 1988 credit and narrow money grew fairly rapidly; inflation increased correspondingly. To finance the growing public sector deficit, credit increases rose, reaching a peak of 7.5 percent of GDP in 1986[20]. However, increases in credit to the private sector fluctuated between 1 and 3 percent of GDP through most of the period (see Table 1.12), which was not enough to keep the level of private credit constant as a share of GDP.[21] The growth in the stock of money (M1) peaked at over 40 percent per year in 1987 (the increase in the nominal stock was equivalent to 2.5 percent of GDP). A credit policy significantly more expansive than increases in M1 and quasi money was made possible by large losses of foreign reserves; between 1981 and 1988, the Central Bank was drained of US$550 million, equivalent to almost 15 percent of GDP in 1986.

1.59 Inappropriate interest rate policies discouraged quasi-monetary savings after 1981. As a result the stock of quasi money fell from 10.8 percent of GDP at the end of 1981 but only 5.6 percent at end 1988. The stock of deposits in foreign currency were especially affected; they were 3.1 percent of GDP at end 1981 and dropped to 0.5 percent in 1988 (though this may partly reflect an inappropriate exchange rate used to value them). Quasi money in local currency also declined from the 1983 peak of 9.7 percent of GDP to just 5.1 percent of GDP in 1988.

1.60 Inflation (defined by the implicit GDP deflator) roughly followed the evolution of the narrow money supply (M1). The low expansion in the money supply and the slow down in the growth of credit between 1980 and 1982 produced a strong reduction in inflation, which dropped to 5 percent in 1982, down from 16.5 percent in 1980/81. But the rapid growth of money and credit that followed, to finance the public sector, accelerated inflation to a peak of about 30 percent per year in 1986/87. The growth of money and credit slowed somewhat in 1988, reflecting not so much prudent credit policies as larger losses of foreign reserves, and inflation declined accordingly. In fact, inflation fell to about 15 percent in the 12 months ending in February 1989 (measured by the consumer price index).

1.61 Conditions changed radically after February of 1989. Domestic credit policy became highly restrictive but narrow money grew faster than in any other year during the eighties. The amount of added credit represented just 1.1 percent of GDP, implying a substantial cut in the real value of the stock of credit. As noted earlier, the public sector became drastically contractionary: there was a corresponding reduction in the nominal stock of net credit directed toward it amounting to 2.1 percent of GDP. The private sector, on the other hand, received more credit, equivalent to 3 percent of GDP. However, this was enough to barely increase the share of private credit in GDP, since inflation exceeded 30 percent. Although credit to the public

---

[20] Including the Central Bank's foreign exchange subsidy to the public sector, which is substantive after 1984. Without this subsidy, credit to the public sector declined (the flow was negative) in 1987 and 1988 (as well as in 1989), and was very small in 1984 and 1986.

[21] Private credit's share was 11.1 percent in 1989, a little higher than in 1988 (10.8 percent) and well below that of 1980-82 (16.2 percent).

sector decreased and the quasi-money stock continued to fall as a percentage of GDP[22], the narrowly defined (M1) money supply increased more than 46 percent due to the massive inflow of foreign exchange that the Central Bank purchased at a fixed rate.

Table 1.12: PARAGUAY - FINANCIAL SYSTEM FLOW OF FUNDS
(Changes in Financial Assets and Liabilities as % of GDP)

	81	82	83	84	85	86	87	88	89	90
M1	0.1	-0.1	1.6	1.7	1.9	1.7	2.5	1.6	2.6	1.6
QM Local Curr	2.4	1.3	1.8	0.8	0.8	1.5	1.1	0.3	1.2	1.0
QM For curr	0.1	-0.7	-0.5	0.0	-0.3	0.3	0.1	0.0	1.9	1.7
<u>Total Liab.</u>	<u>2.6</u>	<u>0.4</u>	<u>2.9</u>	<u>2.5</u>	<u>2.5</u>	<u>3.6</u>	<u>3.7</u>	<u>1.9</u>	<u>5.8</u>	<u>4.3</u>
For Res	0.8	-2.2	-0.7	-2.4	-0.8	-1.1	0.7	-1.6	3.3	4.7
Cred	4.1	2.0	3.0	4.6	4.3	7.5	5.6	4.6	1.1	0.5
Public Sector	1.7	0.8	2.4	2.1	3.2	4.1	3.4	2.2	-2.1	-2.2
Private Sector	2.4	1.2	0.6	2.4	1.1	3.3	2.1	2.4	3.3	2.7
Other	-2.3	0.6	0.6	0.3	-1.0	-2.8	-2.6	-1.1	1.4	-0.9
<u>Memo Items</u>										
% change in M1	0.6	-1.8	22.8	25.7	29.6	27.6	42.9	25.7	46.1	27.7
% change in Pvt. Credit	18.4	7.7	4.4	20.8	10.4	36.9	23.4	27.9	42.2	34.8
Inflation (GDP def.)	16.3	5.0	14.4	26.9	25.2	31.5	30.3	25.1	31.2	36.3

Source: Central Bank of Paraguay and World Bank estimates.

1.62   Similar trends developed in 1990, when a further drop in credit to the public sector was accompanied by a small increase in credit to the private sector (the real value of the stock declined this time). Nonetheless, monetary expansion reached almost 30 percent. Again the main reason for the monetary growth was a massive accumulation of foreign reserves. The Central Bank accumulated almost US$180 million between February and December of 1989 and US$250 million in 1990. With the rise in money growth, inflation accelerated to 31 percent in 1989 and to 44 percent in 1990.

1.63   Monetary policy has been less expansive in 1991, despite substantial accumulation of foreign reserves (US$300 million between the end of 1990 and August 1991) again this year. Stronger public finances explain further reductions in public sector credit; the liberalization of interest rates combined with lower interest rates on foreign currency has helped cut private sector

---

[22]   Quasi-monetary deposits in national currency dropped to 4.9 percent of GDP. Quasi money deposits in foreign currency increased in dollar terms from the low levels of 1988 and in terms of Guarani rose spectacularly because of the change in the exchange rate used to value them.

borrowing locally and shifted it abroad. Another and most encouraging development is taking place in quasi-monetary savings. These savings, both in local and foreign currency denominations, seem to be recovering fast reflecting the higher real interest rate on savings. A progressive reduction in reserve requirements that started in 1991 will strengthen this recovery by further increasing interest rates on savings through cutting the spread between lending and deposit rates. The higher savings have helped cut the rate of growth of the money supply. M1 grew at an annual rate of 23 percent in the first semester of 1991 (there is little seasonal growth in the second semester). Inflation has receded with the slowdown in monetary growth. The annual rate of inflation in the first semester of 1991 was below 15 percent and this figure is not expected to e exceeded in the year as a whole.

CHAPTER II: SECTORAL CONSTRAINTS AND OPPORTUNITIES FOR GROWTH

2.1   Agriculture has been and will likely remain the country's most important growth engine. Constraints on its growth--linked to difficulty in further expanding the agricultural frontier and transportation--will affect the whole economy. As a result of the construction of Itaipu, Paraguay is now exporting large amounts of its surplus electricity to Brazil. The revenues from these sales should become an important source of additional foreign exchange and therefore provide opportunities for growth. Finally, the informal sector has been an important growth source since the mid-1960s. However, this source may decline soon if Argentina and Brazil significantly cut protection for their domestic markets.

A. Agriculture

2.2   Paraguay is rapidly approaching a crucial juncture in the development of its agricultural sector. The various stages of agricultural development through which Paraguay has passed have all, to a greater or lesser degree, have been extractive and/or exploitive of natural resources. With land as an abundant factor, the natural renewal of fertility and the low level of intensity of land use allowed this approach to operate until recently as a sustainable form of production. The acceleration of development over the past 20 years, however, has meant that the limits to extensive growth soon will be reached. There is a potentially serious problem of increasing degradation and erosion of soils in the areas already taken into production. A radical change in the strategic approach to the development of the sector must be adopted, if sustainable growth is to be achieved in the future. This implies a concentration on increasing productivity in tandem with a heightened concern with the environment and the preservation of natural resources. The key issue is how far a traditionally weak public sector has the will or the capacity to influence the course of events through technical or policy-based interventions.

2.3   The activities of product marketing, including concentration, transport, primary processing and distribution or export, are mainly undertaken by the private sector. The role of the state in this area has been limited to the establishment of certain quality norms, the establishment of basic prices between producers and marketing agents (for certain products), the creation of price information systems, and in very limited cases (e.g., wheat, fruit, and vegetables) the creation of marketing infrastructure. Cooperatives are playing an increasingly important role in marketing both for small and medium-sized farmers. Marketing intermediaries have an important function beyond the simple commercial transactions in products, especially in the provision of credit.

Productivity

2.4   The main source of growth in the dominant crop sector has been area expansion, rather than increases in yield. Yield figures, to the extent that they can be believed, show modest but very uneven growth for cotton, wheat and soybeans and little or no growth for maize. This is not a surprising picture in a country where land has not been a constraint and where farmers have therefore had little incentive to intensify their production systems. Such yield increases as there have been in the export crops have almost certainly spilled over from Brazil *via* the introduction of improved varieties of wheat and soybeans developed by the Brazilian research establishment.

2.5   Yields in Paraguay are significantly lower than those achieved in the US, but they are approximately equal to those achieved in Brazil and Argentina. Obviously, there is a large theoretical potential for improved yields derived from increased use of fertilizers. However, there is a wide divergence between theory and practice. The main determinant is not access to technical know-how, but the interplay of various sets of incentives. The most obvious one is pressure on land. Farmers generally do not take risks with high-input agriculture if it is possible to expand production by cultivating a wider area. The intensification of Paraguayan agriculture will not occur therefore on any impressive scale until the limits of the frontier are reached. There are, however, a number of other important factors, the most obvious of which is the ratio of fertilizer costs to output prices. Due to its location, it is unlikely that Paraguay will ever compete in this respect with Brazil, Argentina or the US. However, in time there will be significant opportunities for achieving economies of scale in fertilizer imports, and the best way of realizing these opportunities will be through the abolition of any remaining regulatory impediments to import and distribution.

Agriculture Taxation

2.6   The tax burden on the agricultural sector, resulting from instruments specific to the sector, falls into three main categories: (i) real estate taxes, levied on officially assessed land values; (ii) export taxes, including export duties, taxes on exchange transactions and stamp duties; and (iii) various internal taxes and fees related to specific commodities. The available evidence does not indicate that the tax burden is higher on the agricultural sector than elsewhere in the economy. If it is accepted that the level of taxation in the economy as a whole is low, at between 8-10 percent of GDP, and that an increase in tax revenue is both essential and likely given the government's spending aspirations and the need to eliminate the fiscal and quasi-fiscal deficits, the issue of taxation in the agricultural sector becomes of crucial importance. The main problems of the present system are its complexity, cost, and inequitable nature. These deficiencies should be remedied in a rational tax system, which in addition should be nondistortionary with respect to the allocation of resources, and conducive to capital formation and economic growth. The design and implementation of such a reformed system will, however, be long and arduous and may be a serious limitation on the ability of the public sector to influence the course of sectoral development.

2.7   In the agricultural sector, emphasis should be placed on improvements in the real estate tax, where there are three priority areas for improvement: (1) a more realistic assessment of property values; (ii) a more efficient approach to the identification of taxable properties; and (iii) a more rigorous collection effort. Structural changes in the property tax system would allow its conversion into a land tax based on production potential, which in turn would allow a rationalization of land use and the encouragement of the adoption of technology, _via_ investment, and hence an increase in both output and productivity. At the same time, steps should be taken to remove explicit and "quasi" export taxes (e.g., exchange taxes and stamp duties). The continued use of fees or user charges is appropriate but should be subject to the same rationalization as the rest of the tax system.

Environmental Considerations

2.8   The most important environmental issues in the agricultural sector concern deforestation, natural resource conservation and the survival of the indigenous communities. The rate of deforestation for agricultural purposes is accelerating. It has increased over the last year in step with growing claims by landless people, since land under forest is deemed to be unused or underused and therefore subject to expropriation. The motive force for much deforestation is the insatiable demand in Brazil for timber, especially from the northern part of the Eastern Region with its "frontera seca" which eases smuggling. The demand for wood from the charcoal-fueled ACEPAR steel plant is likely to put pressure of forests in the future. Finally, in many areas, land has been cleared of forest for agricultural and livestock purposes, in some cases using low cost credit, with only an incomplete use of timber and without due regard for the capability of the soils to sustain agriculture over the long term. Although some land-owners are respecting the basic rules concerning slopes, water course protection and contour ploughing, there are many areas where total forest removal and windrowing of debris down the slopes are creating the potential for soil erosion on a massive scale. In some areas, forest clearing has revealed poor soils which, after a few years of mining the natural fertility, are abandoned to forest regrowth.

2.9   In theory, cutting of forest and transport of logs are both subject to permits that should be issued by the Servicio Forestal Nacional (SFN), which comes under the Subsecretaría de Recursos Naturales y Medio Ambiente (RNMA) in MAG. However, the SFN is unable to control forest cutting for lack of resources and because of the significant pressures exerted by landowners and others engaged in exploiting the forest. In any event, a realistic view of the situation suggests that policing and control measures are unlikely to stem a tide that has its origins in the powerful incentive system described above. The only hope for slowing this process lies in: (i) the development of the technical basis for natural resource planning and for soundly-based expropriation; (ii) a change in expropriation policy away from the simplistic view that forested land is by definition underused; and (iii) consolidation and development of productive potential in existing colonized areas to reduce the flow of new migrants seeking land.

2.10   The "agricultural frontier" in Paraguay has traditionally been seen as virtually limitless, whereas in practice deforestation will probably be complete in little more than 10-15 years unless action is taken now to prevent it. Simple, emotional appeals not to cut down trees are unlikely to have any impact, and it is important that the debate should be carried out on the basis of detailed technical knowledge. Cutting down forest to make way for sustainable agricultural production can be a rational use of natural resources where the soils and slopes permit. However, the absence of strategy and technical knowledge leads to indiscriminate deforestation. The generation of information about the natural resource base is essential to allow the definition of strategy and the establishment of mechanisms to implement it. The establishment of a satellite image-based Geographic Information System (GIS) and Agricultural Land Information System (ALIS) are therefore key requirements.

Future Agricultural Development Strategy

2.11   The development potential of the agricultural sector in terms of underutilized resources, available technology and markets is sufficient to allow the sector to continue to play an important role in economic growth and development. However, the "easy" phase of Paraguay's agricultural development is coming to an end. A new set of policy measures and objectives is required to make more productive use of the resources available. Within this context the key issues that must be woven into any future growth strategy are as follows: (i) the limits to growth imposed by the degradation of natural resources; (ii) the potentially explosive impact of landlessness and rural poverty; (iii) the extremely poor data base on which to make rational decisions concerning poverty and natural resource protection; (iv) the weakness of the institutions that will be needed to develop this data base; and (v) the generally high input: output ratio (caused by geographic factors and infrastructural weakness) that discourage the adoption of more intensive systems of farming.

2.12   To deal with these issues implies a fundamental structural change from the present policy of continuous expansion of the agricultural frontier, to a strategy of intensification of resource use through sustainable agricultural practices. The expansion of the agricultural frontier via timber extraction and deforestation is coming to an end. The forest cover is not yet completely gone, but deforestation is accelerating and there are only a few years left. What is absolutely clear is that, given the state of knowledge about land resources and their ownership, there is an urgent need to prepare land information systems to provide the basic data necessary to implement a long-term strategy. The definition of strategy, however, does not need to wait for the compilation of this detailed data base.

2.13   Given the importance of the small farmer and the pioneering nature of much of the agricultural development achieved to date, a significant change in priorities in public sector expenditure, both capital and recurrent, will be required if rational and sustainable growth is to be achieved in future. More specifically, the following expenditures deserve support from external sources:

(a) A Geographic Information System (GIS) and Agricultural Land Information System (ALIS), based on satellite imagery and closely coordinated with a new cadastre, should be established. This would allow the MAG to implement controls on deforestation and to promote practices designed to minimize soil degradation and erosion and to protect watersheds.

(b) The Instituto de Bienestar Rural (IBR) should be reorganized and strengthened, in order to discharge its role of colonization of agricultural areas effectively. The purchase of large holdings for fractioning, either through legal channels of expropriation or through the market, will remain a key role of IBR. Financing measures, including bonds amortized through the payments made by those receiving land under the settlement schemes, must be sought to limit demands on the Government's resources.

(c) There is an important and legitimate role for the government in assisting the agricultural development process through agricultural research, extension, quality and phytosanitary control for exports, control of land clearing and promotion of soil conservation practices. Support should be given to the promotion of farmers' organizations, whose degree of development was greatly restricted under the previous government. These are the most cost-effective means of providing many essential services to the small farmer. Efforts in this regard should be especially focused on the high potential Eastern Region, where the country's main comparative advantage lies.

(d) Increased tax collection in the agricultural sector is needed to finance an enhanced public sector role as described above. A major effort will be required to create a complete cadastre and land titling system, and install a mechanism to keep it up-to-date once created. In taxation, the first phase should be to ensure that all landowners are liable for and actually pay the <u>Impuesto Inmobiliario</u> (estate tax). The cost of issuing the invoice is small, and there are strong arguments for having a minimum taxable land holding of a very small size. Later phases should then aim at increasing the official (fiscal) value of land to its real market level. As a later refinement, a combination of the cadastre plus production, economic and soil capability overlays would yield a tax more closely related to the intrinsic productive capability of the soil (e.g., IMAGRO-Uruguay or the proposed, more sophisticated, version from Argentina) and would be broadly production neutral. Such a tax could be made production positive by including allowance for capital investments made on the land.

B. Population, Colonization and Poverty

2.14   The country's population was decimated after the War of the Triple Alliance in 1865-70. Raine estimates the Paraguayan population at the beginning of the war at 800,000; by the end it was reduced to about 230,000 (see Table 2.1). A significant loss of population also occurred in the Chaco War in the 1930s. Underpopulation typified the country into the late 1950s. Paraguay was then and remains one of the least densely populated countries in Latin America.[23] Thus, for many years the Government encouraged foreign immigration.

Official Emphasis on Foreign Immigration Until the 1940s

2.15   The first foreign colony was settled in Paraguay in May 1855 by French immigrants. It did not succeed and no further attempts at colonization were made until after the War of the Triple Alliance. The first successful permanent colony was founded by German immigrants in San Bernardino in 1881, which later became an almost self-sufficient dairy center. Today, San Bernardino is an important municipality and summer resort. Further attempts were made by Germans and Australians soon after 1881, but they were not successful. Many German colonists arrived after World War I, and most did well (Raine 1956, p. 130).

2.16   A difficult but successful colonization enterprise was that by the Mennonites in 1926, the only foreign settlers in the Chaco--even though severe droughts in 1937 forced the relocation of almost 20 percent of these settlers to the more fertile eastern territories. Their practice of hard work, proper planning, and adequate financing provides a lesson on how to prosper in a hostile environment. Though not from Germany (the first group came from Canada, another from Russia), most Mennonite immigrants were of German descent. The most serious obstacle they faced in the Chaco was lack of transport infrastructure. Despite the difficulties, shortly after World War II, these colonists were raising beef and dairy cattle, producing cotton, peanuts and beans, and building refrigeration plants. In the early 1950s, the Chaco supported 9,000 Mennonites, about 5,000 of whom were living in eastern Paraguay (<u>Ibid.</u>, p. 305). More immigrants arrived just before and after World War II. A Japanese colony was established in 1936 about 50 miles from Asuncion near a rail line; by 1946 it comprised more than 600 people.

---

[23]   In 1987, there were 10 people per square mile of territory, and 22 people per square mile of agricultural land (World Bank, <u>Social Indicators of Development, 1988</u>).

Table 2.1: PARAGUAY - ESTIMATED POPULATION 1865-1950
(Selected Years)

Years	Population (thousands)	
	A. Raines	B. Mendoza
1865	800	600
1872	231	231
1887	329	328
1899	635	430
1909	650	541
1919	800	683
1938	950	1062
1945	1100	1247
1950	1405	1397

Source: Raines 1956, p. 295; Mendoza, in Rivarola and Heisecke (eds.) 1970, pp. 17, 21.

2.17   Spontaneous (not officially sponsored) Brazilian immigration was of special importance in the 1960s and 1970s. There are few statistics that gauge the size of this inflow. Incomplete records suggest a lower bound of 6,000 Brazilians a year entering the country in 1974/75. Birch quotes estimates of 300,000 Brazilian colonists in 1979 and 420,000 in 1982 (the latter figure apparently provided by a Brazilian official in the eastern border region (Birch, n.d., p. 31).[24] Several factors made this immigration attractive: (i) the rapid colonization of Parana in Brazil had ended and the fertile virgin land was rapidly being exhausted; (ii) nearby virgin lands in eastern Paraguay had identical soil and climatological characteristics; (iii) despite lack of transportation infrastructure, land prices were much lower in Paraguay than in Brazil; (iv) taxation in Paraguay was much lower than in Brazil; and (v) communication with Brazil and the rest of the world was relatively easy, as it was easy to reach the Atlantic given Brazilian transport infrastructure and the port privileges offered to Paraguay in Paranagua.

---

[24]   For population purposes, it is also important to know Paraguayan emigration to neighboring countries, but little information is available (see Mellon and Silvero in Rivarola 1970, p. 53). Some authors suggest that for political reasons, as well as internal strife, emigration could be significant. Others disagree. Eligio Ayala, in "Migraciones" (Santiago, Chile, 1941) holds the view that this migration has been important; Rivarola, however, states that its effect was not important, at least before 1970 (in Rivarola 1970, p. 32). The only official figure mentioned is from the 1947 Argentine Census, which states there were no more than 50,000 Paraguayans living in Argentina (in Rivarola 1970, p. 41). The Paraguayan boom during the seventies and the far bigger economic slump in Brazil and Argentina than in Paraguay in the eighties suggests that migrations from Paraguay would not have become more intense after the 1970s.

Colonization to Fight Poverty After 1940

2.18   The colonization program's main concern changed after 1940 to poverty alleviation for the local population, which new goal was embedded in the Agrarian Law of 1940. The population was concentrated around Asuncion, where minifundia was the prevailing land tenure system[25]. Before 1940, most colonies were settled by foreigners; after 1940 most official settlements were comprised almost exclusively of Paraguayans (Raine 1956, p. 259). In practice, however, the impact of the Agrarian Law was not significant. The forties and fifties produced little progress in alleviating poverty and minifundia remained abundant around the capital.

2.19   Spontaneous colonization of the eastern region began in those years, but without sufficient official backing, Paraguayan settlers did not prosper. This led the Government to give high priority to solving the problems posed by the minifundia area around Asuncion. The Instituto de Bienestar Rural (IBR) was founded with the objective of relieving population concentration in these areas, legalizing tenancy of the settlers occupying Government or private lands through colonization programs, and providing technical assistance, credit, and infrastructure support to settlers in the colonies (WB 1984a, p. 9).

2.20   IBR had a powerful effect. From 1963 to 1982 its settlements had covered 5.6 million ha on 72,000 individual lots. In addition, 73 settlements on 21,000 lots were established on private lands covering 840,000 ha[26]. The size of public properties has been subject to conflicting reports and land titling is an important and unresolved issue. Raine reported that at the end of the nineteenth century hardly any land remained in public hands (see Appendix I para. 17), but most IBR's colonies were settled on public lands in the 1960s and 1970s. As for legalization of titles, there was not much improvement in relative terms--50 percent of all farms had definitive or provisional titles in 1956 versus 57 percent in 1982--but the number of farms almost doubled in that period rising from 150,000 to 245,000 (WB 1984a, p. 10 and Table 14). A feature of these programs was that most beneficiaries actually paid nothing for their lots, even though prices charged were well below market values. But as most settlements were on state lands, this failure did not drain IBR's financial resources.

2.21   Despite often poor sanitary conditions, Paraguayans are among the better fed peoples of the Western Hemisphere (Raine 279, Table p. 280). In the countryside, meat, not bread has been the staff of life, with daily per capita consumption of about 200 grams and an intake of 3,200 calories.[27] The lower class city dwellers are better off than most Latin Americans and perhaps better off than persons of a similar class in many parts of the world; however, they do not eat as well as farmers, averaging 2,800 calories per day (Ibid., p. 288).

---

[25]   Despite all the laws enacted and resources spent, the agrarian problem remained urgent and was a cause of deep concern, according to Eligio Ayala, a Paraguayan scholar writing in 1941 (quoted in Rivarola 1979, p. 31).

[26]   More recent estimates, though not necessarily comparable, put the number of lots on official lands at around 125,000 and on private lands at 60,000 ("World Bank Environmental Issues Paper," June 1990, p. 3). The former figure is at variance with a forecast that claimed that State-owned lands were exhausted and further settlements would require purchases of private property (WB 1984a).

[27]   Estimates based on a study of a small town, Pirebebuy, in the Cordillera Department (minifundia region) (Reh, E., "Paraguayan Rural Life," Washington, Institute of International Affairs, 1946, quoted in Raine 1956, p. 287).

Difficulties Ahead

2.22   Despite these achievements, the agrarian problem remains urgent and causes deep concern. The recently acquired freedom of expression in Paraguay has presented the new administration with a series of challenges, the most important of which is a vociferous and growing demand on the part of campesinos for land. Land occupations, confrontations, and evictions are becoming more common and more serious. The government must deal tactically with this issue, while developing a long-term strategy for the small farm sector. With still extensive cultivable land resources and a low rural population density, the problem would seem to be relatively trivial. However, the distribution of land is very skewed and worsened significantly over the past 35 years, as sizeable areas of tierras fiscales or state-owned land were distributed to domestic and foreign friends of the previous government.

2.23   The land distribution policies for small farmers, established in the early 1960s, have been only partly successful. Many families have been settled on land, but the institutional arrangements have been seriously deficient. The Instituto de Bienestar Rural (IBR) has given away large areas of state-owned land at little or no cost but has failed to provide many small farmers with titles or to collect payment for land distributed. The net result is that: (a) the remaining area of tierras fiscales actually are available to the government for distribution to small farmers is unknown but probably quite small; and (b) the amount of tierras fiscales distributed en forma prebendaria and which, in theory at least, might be susceptible to "expropriation" or other coerced return to state ownership, is similarly unknown. To this can be added that the true number of landless families and the farmers who possess land without legal title, and therefore are without access to formal credit, which is also an unknown number but probably quite large.

C. Transport

2.24   Lack of transport infrastructure has been a major bottleneck affecting the country's foreign trade and efficient use of resources through most of Paraguay's history. The most important international means of communication used to be the waterways. There are conflicting reviews on the quality and costs of the services rendered. Raine wrote in the mid-1950s that river service was said to be good, provided by modern vessels that have run regularly for a long time; but he complained that river transport was an Argentine monopoly and that it cost more to send goods from Asuncion to Buenos Aires than to ship from Buenos Aires to the United States, Europe, or the Far East, with freight costs downstream double those upstream (Raine 1956, p. 383). However, a Bank report published about the same time takes an opposite view: "most ships and barges now operating on the Parana and Paraguay Rivers are overaged and ill adapted for river transportation" (WB 1959, Annex II, p. 2); moreover, "analysis of freight rates does not confirm the widely held view that transportation costs on the Paraguay and Parana Rivers are excessively high" (Ibid., Annex II, p. 2).

2.25   At present, river transport costs to Atlantic ports on the River Plate Bay are lower than land transport costs to Paranagua, Brazil, a deeper water port on the Atlantic at the same latitude as Asuncion. However, because Paranagua accepts large ships offering lower freight rates to ports in Europe or the US, overall it is less expensive to reach these ports by land through Paranagua than by river using River Plate ports. Thus, the fact is that after the opening of roads connecting to Brazil, riverways increasingly became commercially obsolete. In 1988 for example, 98 percent of all freight and passenger transport services was provided by roads (see Table 2.2).

Table 2.2: PARAGUAY - TRANSPORT SERVICES 1986

	Freight (ton-km)a/	Percentage	Passeng. (Pass-km)a/	Percentage
Roads	6356.0	97.6	2698.7	98.4
River	114.5	1.81	1.5	0.4
Railroad	19.0	0.32	6.2	1.0
Air	0.0	0.0	6.7	0.2
TOTAL	6489.5	100.0	2743.1	100.0

Source: Plan Nacional de Transporte 1988-92.

a/ in millions.

2.26 Regarding domestic transport facilities, there is agreement that transport development lagged for a long time and that until the 1960s many areas of the country were inaccessible to wheeled traffic, or could be reached only under the most favorable weather conditions (Ibid., p. 1).

2.27 The Bank report quoted above asserts that there are few reasons for the bottlenecks posed by the sector. The lower sections of the Alto Parana and the Paraguay Rivers are often claimed to be the best unimproved rivers for inland navigation in the world. Regarding roads, the report maintains that there are few countries where topography lends itself so easily to road construction. Because terrain is flat and easy, road construction costs are low as are vehicle operating costs. Moreover, in some areas first-class road construction material is readily available (Ibid., p. 2). Since the late 1930s, the Government has been determined to reduce transport dependency for riverways and to expand the network of roads. It did so by substantially increasing resources for investments in the sector and by reaching important agreements with Brazil.

Main Agreements with Brazil

2.28 In 1939, General Estigarribia, newly elected President of Paraguay, went to Brazil and signed the first of a series of agreements that drastically changed the structure of the transport system in Paraguay. The accord was for the construction of a spur off the Sao Paulo-Campo Grande railroad to Ponta Pora on the Paraguayan border. This spur was expected to connect with the projected Concepcion-Pedro Juan Caballero rail line in Paraguay, thus allowing Paraguay an alternative other than the river for access to the Atlantic Ocean. An option then available was the Central Paraguayan Railroad, which only went south from Asuncion to Encarnacion, linking the latter to the Argentine system (and thus to Buenos Aires) after crossing the Parana River (Birch, undated p. 7) by ferry.

2.29   In 1941, Paraguay and Brazil signed several other accords. In one, Brazil gave Paraguay free facilities at the Atlantic port of Santos, close to Sao Paulo; another stipulated the conditions for the construction and a 30-year concession for the operation of the abovementioned railroad between Concepcion and Pedro Juan Caballero. A Brazilian firm was awarded the contract to extend the lines from the Ponta Pora spur to the new line. This accord was revised in 1946, but little work was done within Paraguayan's borders; however, the Brazilian side was completed in 1954 (Ibid., pp. 9-13). Another accord (signed in February 1957) provided the basis for the construction of a road between Concepcion and Pedro Juan Caballero--also important for Brazil to help develop its own areas near Pedro Juan Caballero (Ibid., p. 20).

2.30   A treaty signed in 1954 secured Brazilian financing for the construction of the road from Colonel Oviedo to the Parana River (now Ciudad del Este) by reactivating an old loan made by Brazil to Paraguay in 1942; works began in 1955 (Ibid., pp. 9-13). In January 1956, an agreement was signed whereby Paraguay gave Brazil free port privileges at Concepcion on the Paraguay River, while Brazil gave Paraguay similar privileges at Paranagua: a port on the Atlantic, south of Santos, directly east and connected to Asuncion by the Colonel Oviedo Road, which would link Paraguay with the Brazilian highway system going to Paranagua (which Brazil completed in the late 1950s). In this accord Brazil also agreed to finance the construction of a bridge across the Parana River. Traffic across the Parana was minimal at the time of the accord. The bridge was completed and inaugurated in 1965 and Asuncion was then connected to the Atlantic by an excellent highway. This was a critical ingredient for the massive increase in trade with Brazil that developed in the seventies.

Evolution of the Road Network

2.31   The road network has increased dramatically since the early 1940s (see Table 2.3). The network multiplied more than four times in 15 years between 1940 and 1955 (a 10.3 percent year increase), and almost doubled again between 1955 and 1960 (13.3 percent year increase). Especially remarkable was the expansion of paved roads (14.8 percent per year). Earth roads also showed an important though less impressive increase. Initially, most paved roads began in Asuncion and went east to Colonel Oviedo, and southeast to Encarnacion, bordering Argentina. Roads were later extended to Iguazu, then to Ciudad del Este, bordering Brazil; Colonel Oviedo was later connected to Encarnacion.

2.32   In the first half of the 1960s, road expansion was as fast as in the latter half of the 1950s (15.6 percent per year), with the extension of paved and earth roads more than doubling. The pace slowed to less than 10 percent per year in the second half of the 1960s, and even more so in the 1970s and 1980s, but paved roads continued to increase faster than the rest. In 1966 the fully paved route to Ciudad del Este (on the Brazilian border) was inaugurated and later upgraded. Other important roads are the route to Bolivia across the Chaco, paved only half way at present; the Concepcion-Pedro Juan Caballero Road (linking the Paraguay River with the Brazilian border at Ponta Pora), also partly paved, and a road linking Encarnacion and Ciudad del Este on the eastern side of the country, which is now fully paved.

Table 2.3: PARAGUAY - ROAD NETWORK
(in km)

Years	Earth	Gravel	Paved	Total
1940	69.4	188.0	12.0	269.4
1945	255.0	354.8	84.5	694.3
1950	288.4	474.4	88.0	850.8
1955	557.1	513.8	95.0	1,165.9
1980	1,317.8	653.2	194.5	2,165.9
1960	2,245.8	653.2	194.5	3,093.5
1965	4,965.3	963.5	470.0	6,398.8
1970	8,012.4	594.4	817.0	9,423.8
1975	9,991.1	582.0	905.0	11,478.1
1980	14,724.6	482.9	1,469.4	16,676.9
1985	20,479.6	452.0	2,076.6	23,008.2

Source: For 1940-60 from Ugarte 1983, p. 193; for 1960-85 from "Plan Nacional de Transporte 1987-1991," Ministerio de Obras Publicas y Comunicaciones (earth roads in MOP figures include roads built by IBR (Instituto de Bienestar Rural), local "Juntas Viales," and private institutions (mennonites, oil companies, etc.), which Ugarte figures do not).

2.33　　The results of the expansion in the road network were impressive. Not only did the country become more independent of river transport and an outdated railway (connecting Asuncion with Encarnacion), but the expanded agriculture frontier could be serviced, and trade with a fast developing neighboring region in Brazil was greatly facilitated. Colonel Oviedo traffic experienced a phenomenal growth in the 1940s and 1950s due to the hinterlands that the highway to Ciudad del Este opened (Raine 1956, p. 311). Ciudad del Este followed suit when roads to it were completed and construction of Itaipu began. The growth of Colonel Oviedo, Ciudad del Este, and the eastern agricultural region illustrates how economic development and the production of wealth follows the building of roads into undeveloped regions. It remains to be seen if opening roads into the untapped Chaco will have similar beneficial effects.

Challenges Ahead

2.34　　Despite many achievements, the transport sector still faces important constraints. The road network has reduced the role of riverways, even for bulky products going abroad such as soybeans and cotton (now the country's two main exports). However, it is not clear that the present lack of river transport represents economic inefficiency; instead this may be the result of poor investments by the State (FLOMERES, the river transport company) regulations hindering private investment, and inefficiencies in the River Plate ports. Port costs in Argentina and

Uruguay may drop if the privatization projects proceed as planned. The Bank is presently carrying out a study of transport costs for exports under different alternatives. If river transport turns out to be less expensive under the new circumstances, Paraguay will need to improve its ports and the navigability of its rivers. A working international accord that would preserve the navigability conditions of these rivers inside and outside the national territory would be extremely useful for Paraguay and all the countries in the River Plate Region.

2.35   Important roads planned sometime ago were built (Concepcion-Pedro Juan Caballero and the Trans-Chaco Highway), but their paving has been delayed. Maintenance also seems a problem; the Bank recommended in 1984 that the Government pay more attention to improving maintenance and upgrading rural roads (WB 1980, p. iii). In addition, public funds are especially scarce now and may remain so for some time. Thus, the Government faces difficult choices: improvement of existing highways such as the two just mentioned; maintenance of the existing network; or building new penetration roads to open up new areas to create added wealth and employment opportunities.

2.36   The railway is a source of concern. One of the first system in Latin America, now linking the cities of Asuncion and Encarnacion, it has deteriorated to the point where it now is of almost no social or commercial significance. The recently inaugurated bridge over the Parana River between Encarnacion and Posadas (Argentina) connects the Paraguayan railroad with the Argentine system (and therefore with ports in the Atlantic). This development may turn the Paraguayan railway into a profitable endeavor. The question turns on the best policy option: make minimum investments and let the railway slowly die; upgrade it drastically; or build a new system. The study mentioned above will also analyze these topics and should help find an answer to this question. Another line planned between Concepcion and Pedro Juan Caballero and linking northern Paraguay with the Brazilian railways system was never built. At present, it may no be wise to advance further on this idea.

D. Itaipu and Yacireta: Electricity as an Export

2.37   Paraguay has abundant energy wealth in the form of surplus hydroelectric resources that can be exported as electricity to neighboring countries. It began tapping these resources in the 1970s with the construction of the main civil works at Itaipu (the largest hydroelectric power plant in the world). Recently, Paraguay has begun to profit from the investment done in the 1970s. Itaipu, a binational entity owned in equal shares by Paraguay and Brazil, has a power capacity of 12,600MW based on 18 turbines delivering 700MW each. Most of the plant is in operation; the last turbine is expected to be installed in early 1991. With all turbines in operation, Itaipu is expected to produce 75,000GWh/year, and half of Itaipu belongs to Paraguay. Paraguayan energy demand in 1989 was estimated at just 1,950GWh/year. Most of this demand is met by Acaray, another hydroelectric plant used for domestic consumption. Its initial power was 90MW, later expanded to 194MW. Therefore, most of Paraguay's share in Itaipu's electricity is sold to Brazil and should represent a continuing source of income for Paraguay.

2.38   In addition, Paraguay participates in another binational entity in the power sector, Yacireta, shared equally with Argentina. Like Itaipu, the agreement that gave life to this entity was signed in 1973, however Yacireta is expected to begin operation only in 1993. Multiple financial and legal problems have delayed its completion. The power of Yacireta will be 2,700MW. Another hydroelectric plant, Corpus, also shared equally between Argentina and Paraguay, is being studied for the long term. Its capacity would be 4,400MW. All energy belonging to Paraguay from these plants would be exported to Argentina.

2.39 As discussed earlier in this report, the construction of these plants (mainly Itaipu) generated a strong boom in the 1970s. Table 2.4 shows the inflow of foreign resources. These resources peaked in the 1978-81 period when they reached close to 10 percent of GDP (nearly US$500 million in 1981), and dropped thereafter, reaching less than 3 percent of GDP in 1986 (barely US$90 million). It was expected that Itaipu would begin operations in 1983 and would have all its power available in 1988; instead, operations began in 1985, with the last turbine expected to be installed in early 1991. As a result of these delays and of economic difficulties in Brazil, Paraguay has not received the expected operating revenues. As a consequence, Itaipu accumulated arrears with Paraguay, which began to be repaid in December 1989 (US$51 million) and in 1990 (US$106 million).

2.40 Because Yacireta is not yet operational, this section emphasizes foreign exchange earnings from Itaipu only. From Yacireta, Paraguay receives income only from payments to Paraguayan workers employed at the plant and from goods sold to it, about US$60 million a year (see Table 2.4). This figure is not included in the analysis of Itaipu that follows. On the other hand, Paraguay consumes some energy from Itaipu. It has to pay about US$45 million for that energy and the interest and amortization of the credit obtained to pay its initial capital contribution (Canese 1990, p. 64); this outflow is not considered either. In effect, the report conservatively assumes that these two elements cancel each other out. In the future, Paraguayan consumption from Itaipu will grow, but Yacireta will start producing electricity that Paraguay will be able to sell; therefore, the net effect of not including Yacireta is to underproject the inflows of foreign exchange Paraguay will receive in the future.

Table 2.4: PARAGUAY - ITAIPU AND YACIRETA, INFLOWS OF FOREIGN EXCHANGE
(million US$)a/

	Itaipu			Yacireta		Total/
	Constr.	Operation	Total	Constr.	Total	GDP (%)
1974	1.5	0.0	1.5	0.0	1.5	0.1
1975	43.0	0.0	43.0	2.6	45.6	3.0
1976	71.9	0.0	71.9	4.8	76.7	4.5
1977	142.9	0.0	142.9	6.1	149.0	7.1
1978	228.9	0.0	228.9	23.2	252.1	9.8
1979	231.0	0.0	231.0	65.3	296.3	8.7
1980	276.6	0.0	276.6	108.5	385.1	8.7
1981	319.6	0.0	319.6	151.0	470.6	8.4
1982	297.2	0.0	297.2	48.4	345.6	6.4
1983	280.9	0.0	280.9	34.3	315.2	5.6
1984	164.4	0.0	164.4	48.8	213.2	4.9
1985	110.0	5.5	115.5	28.3	143.8	4.5
1986	54.9	7.7	62.6	35.6	98.2	2.8
1987	50.0	9.2	59.2	85.5	144.7	3.9
1988	37.0	9.8	46.8	55.0	101.8	2.6
1989	30.0	63.7	93.7	60.0	153.7	3.7

Source: R. Canese, "La Problematica de Itaipu," Ed. Base-Ecta, Asuncion, Paraguay, 1990 (Tables 1 and 9).

a/ Figures in this table are similar but not exactly the same as the latest balance of payments data.

2.41   The cumulative expenditure on Itaipu at the end of 1989 was about US$18 billion. As Table 2.5 indicates, the project was financed mainly through borrowing; capital is an insignificant fraction. As noted above, the loans were guaranteed by the Brazilian Government. As is normal in these kinds of projects, interest charges represent a significant fraction of total costs. But at 40 percent, they were especially high in Itaipu. Also total cost greatly exceeded initial estimates. When the project started in 1973, the cost was expected to be US$3.44 billion (Ibid., p. 67). Several factors help explain the difference. First, between 1973 and 1989, US prices (index for industrial goods) increased 180 percent, a much higher annual inflation rate than in the 1960s, when the project was evaluated. This explains an added cost of US$6.1 billion, a large fraction of which is hidden in interest charges[28]. Second, the real interest rate may be three percentage points higher than initially forecast, and works lasted two years longer than originally estimated. In a project so "interest rate intensive" as Itaipu, this could explain an additional cost of US$4 billion. Third, the plant was expanded 30 percent (scheduled to have a power of 9,800MW with 14 turbines; instead, 18 turbines will be installed with a power of 12,600MW); at 1989 prices this would be equivalent to US$2.9 billion. And fourth, housing investment was drastically underestimated in the initial project, and in 1989 this represented 15 percent of the value of main fixed assets; in 1989 prices this would be equivalent to US$1.4 billion.

Table 2.5: PARAGUAY - ITAIPU EXPENDITURES AND FINANCING 1984-88
(million US$)

	1984	1985	1986	1987	1988
Expenditures (Cummulative)	11755	13194	14597	16206	17739
Main Fixed Assets (dam)	4300	4504	4822	5157	5509
Housing & Related	668	679	714	764	840
Financial Costs	4471	5452	6205	6879	7562
Liquid Assets (Purchase)	123	220	236	472	658
Other	2193	2339	2620	2934	3170
Financing	11755	13194	14597	16206	17739
Long Term	7750	8494	8375	9740	9301
Short Term	1482	2210	3177	4018	6162
Exch Rate Diff.	2423	2390	2945	2348	2176
Capital	100	100	100	100	100

Source: R. Canese, "La Problematica de Itaipu," Ed. Base-Ecta, Asuncion, Paraguay, 1990 (Tables 17 and 21).

---

[28]   Assets reflect dollar values at purchase time and are not adjusted by dollar inflation thereafter. However in later years, the much higher interest rate (reflecting inflation) adds substantially to total costs.

2.42   For Paraguay, revenues from Itaipu stem from four sources: (i) a 12 percent return on the capital that Paraguay provided at the beginning (US$50 million, as much as Brazil); (ii) the compensation to ANDE (the Paraguayan electricity company) for the administrative costs it incurs in running Itaipu (both (i) and (ii) are received by ANDE); (iii) "royalties" paid for the use of the water from the river; and (iv) sales to Brazil of the surplus electricity that Paraguay does not use. The Finance Ministry receives (iii) and (iv). The royalties and the price to be received by Paraguay for the electricity Itaipu sells to Brazil on its behalf were fixed by the Itaipu Treaty: the first was set at US$650 per GWh generated, and the second at US$300 per Gwh sold. The latter two figures are increased by two factors: first, a multiplicative coefficient set at 3.5 in 1986 and increasing to 4.0 in 1992 and afterwards (Ibid., p. 171); and second, the increase in the price level in the US (a simple average of US consumer prices and prices of industrial goods).

2.43   In 1991, Itaipu should be producing energy at full capacity (about 75,000 Gwh/year, assuming losses of about one-third of potential power). Paraguay is entitled to half of it but almost all is sold to Brazil. As US prices can be expected to be 25 percent higher than in 1986, the adjustment factor would be about 4.9 for 1991. This means that from royalties plus sales to Brazil, Paraguay's Finance Ministry would receive about US$4,655 per Gwh, or US$175 million. Adding the return on capital (US$6 million) and the compensation to ANDE (about US$9 million), Paraguay should receive a total of about US$190 million from Itaipu from these four sources in 1992 and thereafter.

2.44   Until 1992 Paraguay agreed to receive in cash only a fraction of the royalties and compensation to which it is entitled from Itaipu. The rest was taken in the form of a bond amortized in 10 years beginning in 1992, yielding an interest rate equivalent to the average interest rate on loans contracted by Itaipu. Table 2.6 shows the fraction of the receipts Paraguay has agreed to loan back to Itaipu. Because of these loans, Paraguay should receive in cash about US$170 million in 1991 (plus the interest on the roughly US$190 million of loans outstanding).[29] From 1992 on, it should receive not only the full US$190 million annual income from Itaipu, described above, but also 10 percent of the loans granted (the amortization payments on the debt accumulated) plus interest (assuming Brazil remains current on its payments).

Table 2.6: PARAGUAY - CREDITS FROM PARAGUAY TO ITAIPU
(% of royalties and value of sales to Brazil)

1985	100.0
1986	71.4
1987	58.1
1988	45.3
1989	33.1
1990	21.5
1991	10.2
1992	0.0

Source: Canese 1990, p. 173.

---

[29] In addition, Itaipu accumulated arrears with Paraguay, part of which were paid in 1989 and 1990 (see para. 2.39). As of end 1990 the arrears were equivalent to the normal payments expected to have been made in 1990. If paid in 1991, the resources Paraguay would receive from Itaipu would exceed the US$170 million mentioned.

E. Informal Trade

2.45   Informal transactions appear to be significant in Paraguay, which helps to explain the fast growth of the commerce sector. Data on the informal trade sector are based on rough estimates where available. These kinds of transactions take place both in domestic and foreign dealings of many small and large enterprises, and mostly result from the desire to evade taxes. The development of these activities is a cause of concern because such enterprises do not pay their fair share in financing Government expenses. Most important seems to be unregistered imports and exports. These transactions also have a positive side: they set a limit on distortions in relative prices of tradeable goods, and therefore help avoid the massive distortions that have been so common in Latin America. Unregistered trade appears to be large enough so that variations in registered import and export statistics often are of little utility. For example, in 1988, this report's import and export estimates exceed official registered trade by 110.2 and 62.0 percent, respectively. However, the unification of the exchange rate has reduced the incentives for unregistered trade significantly.

2.46   This kind of trade traditionally has been important in Paraguay. There are miles of "dry borders" impossible to monitor, great demand for high-quality imports not available in neighboring countries, high local tariffs and taxes, and cumbersome legal procedures. Paraguay has tried to encourage the legalization of such transactions with only partial success. By 1956 Paraguay and Brazil had signed a General Treaty on Trade and Investment that allowed trade between the two countries to take place outside existing regulations governing trade with other countries--the intention being to reduce illegal trade in border areas. Contraband along the border with Matto Grosso was estimated at 70 to 80 million cruzeiros in the mid-1950s (Birch, n.d., p. 20, from an article in "O Estado de Sao Paulo," January 1956).

2.47   The construction of a bridge over the Parana River at Ciudad del Este, completed in 1965, fostered growth of commercial activities oriented to satisfy luxury demand from Brazilian tourists who faced high tariffs in their country. An upsurge in unregistered imports was detected (WB 1971, p. 57), and the Government contemplated reducing import tax rates on luxury consumer goods (e.g., alcoholic beverages and electronics) to reduce this kind of trade. The wide disparity between statutory and effective taxation left ample room for cutting rates without sacrificing revenues (WB 1971, p. 63). This policy was implemented only at the end of the eighties.

2.48   Paraguay serves as commercial intermediary for imports that actually are headed mostly to Brazil and Argentina (the tourist regime). Though a significant percentage of these imports are included as such in the country's formal accounts, the corresponding exports are unaccounted for, which effectively biases the country's balance-of-payments accounts. Such transactions are an outcome of the heavy protectionist policies prevailing in these bordering countries, and are a cause of concern because of poor prospects for Paraguay if Brazil and Argentina succeed in opening up their foreign trade.

2.49   Other informal activities on foreign transactions involve the so called "triangulation" process. Often these are formal exports from Paraguay that reflect not Paraguay's but a neighbor's output. This behavior is explained by Paraguay's low export taxes, and free exchange rate (a parallel rate openly tolerated before 1989 and the only rate since then), versus heavy implicit (and sometimes explicit) export taxes, pricing policies and often unrealistically low exchange rates in neighboring countries. Paraguay's soybean exports reflecting Brazilian output have been most relevant recently, but coffee exports have also been important in this regard. Argentine meat exports to Brazil also are conveyed through Paraguay when there is a significant difference between the official and free exchange rate in that country.

2.50   A reduction in protection in the context of MERCOSUR may have a severe impact in Paraguay. Paraguay's informal sector has developed to provided Argentinean and Brazilian imported consumer goods unavailable in those countries. With MERCOSUR, these goods will become legally available in all signatory countries, thus the services presently provided by Paraguay will no longer be in demand. In this case, Paraguay will need to develop alternative productive activities to absorb resources now used in the informal sector.

CHAPTER III: POLICY REFORMS TO STIMULATE GROWTH

3.1 Paraguay faces significant structural problems. Some of them affect productive sectors, such as those linked to the alternative of expanding the land frontier versus increasing productivity in the agriculture sector. Solving them will require a long, sustained, and carefully planned effort. Others refer to the policy environment becoming an "artificial" hurdle to the development process. The latter's removal is relatively easier and will facilitate and increase growth.

3.2 In the private sector, the financial sector and trade policy would benefit from reform. In the financial sector, there is a clear need to: (i) make more uniform the regulations affecting different financial intermediaries to avoid artificial specialization among them; (ii) rationalize the functions of the Superintendency of Banks, which because of too many unnecessary responsibilities, does not pay sufficient attention to its most critical role, that of judging the financial health of commercial banks; and (iii) redefine the role of official banks. To these reforms should be added the improvement of the rediscount mechanism and, as the macroeconomic situation permits, the reduction of the reserve requirements to cut the spread between lending and deposit rates. This package should increase financial savings, which were cut in half (as a percentage of GDP) after several years of negative real interest rates.

3.3 In the trade area, tariffs should be restructured to remove the existing excessive tariffs ("water"), discourage informal activities and cut protection. Customs is another structural area where reforms are critical. The country is already effectively operating with low import duties; trade-related distortions, while not important now, may quickly become so. The Customs Code contains high tariff rates that are not being applied--to the good fortune of the country. However, since the code exists, future Authorities may be tempted to apply it. It would be prudent to establish a code reflecting the rates effectively paid currently.

3.4 In the public sector, key problems are the public enterprises and tax reform. Public enterprises made some improvements in their financial performance in 1989, and raised tariffs in 1990, but much still needs to be done. Overdimensioned investments, overstaffing, and lack of management accountability have been sources of difficulties in the recent past. Investment carried out by public enterprises contracted after 1989, so the first problem now seems less urgent. Several enterprises are performing audits, and studies are being conducted on how to improve use of resources, but further efforts are required. Privatization and/or joint ventures with the private sector are being discussed, but these policy overtures have not gone very far.

3.5 The tax system is another area that needs urgent action. The Government already has started taking important actions with a significant effect on tax administration, but the reform of the tax system as such is still to be approved. Evasion has been the main problem. It is rooted in a system perceived as unfair by the private sector because of high tax rates linked to widespread exemptions. For a tax reform to succeed, it will be necessary to have the backing of the private sector--which must perceive it as a better alternative to what they now have. Wider bases, fewer exemptions, and lower rates are the basic ingredients of such a reform.

A. Financial Sector Reform

3.6   Two areas are of special relevance for policy reform in the financial sector: monetary management and institutional reforms. The Government is taking important correcting steps in both. With regards to monetary management, the IMF and the Bank have supported a program of structural reforms that have sought to replace a rigid system of monetary control--which has led to a marked disintermediation from the banking system--to a market based system of control. Furthermore, Paraguay has adopted an ambitious monetary program. In addition, a unified and to some extent free exchange rate policy now in place has critically important implications for monetary policy management. Regarding institutional reforms, Authorities are in the process of reforming the Central Bank and the Banking Laws, and are studying ways to improve banking supervision.

3.7   **Overview of the Sector**. The financial sector's main components are commercial banks, investment banks, official banks, savings and loan associations, and finance companies. Although these institutions are subject to quite different regulations, actual differences in their lines of business are far less significant. However, as a result of inappropriate regulations and inefficient supervision, Paraguay's financial sector is artificially segmented (overspecialized institutions for the size of the market) and fragmented in almost every category (too many institutions in each category). Unregulated "informal" or "parallel" institutions are run by most formal financial intermediaries to avoid existing regulations.

3.8   The country has in operation 23 private commercial banks that mobilize mainly demand and time deposits, including deposits denominated in foreign currency. Foreign banks have a strong presence, with 14 institutions mobilizing more than 70 percent of commercial bank deposits. There are also four official banks in addition to the Central Bank: the Fondo Ganadero, basically a second-tier institution lending to the livestock sector; the workers' (BNT) and development (BNF) banks, which are both first-tier (though not very successfully) and second-tier institutions; and the housing bank (BNV), which channels funds to that sector and supervises saving and loan associations. In addition, there are six savings and loan associations that mobilize mainly short-term time deposits and grant short-term credit (not always to the housing sector), and 28 finance companies that mobilize time deposits denominated in domestic currency.

3.9   Table 3.1 summarizes some indices of the relative size of these intermediaries in selected years. Although BNF remains an important source of credit, its share of credit has declined since the mid-1970s. Most of the credit BNF grants is not financed by deposits, but external credits and loans from the Central Bank (which appears as other net in Table 3.1). Commercial banks (including the official BNT) have maintained their share in domestic resource mobilization and (with some fluctuations) also their share in total credit granted by the system. Private development banks are insignificant and capture virtually no financial savings. Finance companies rapidly developed between 1975-85 but stagnated afterwards; they grant less than 10 percent of the credit commercial banks do and capture an even lower proportion of domestic savings. However, the size of these institutions may be underestimated because of the importance of the "informal" component in them (twice the size of the "formal" component according to unofficial estimates).

3.10   **Supervisory and Regulatory Practices.** In the financial system, economic-type regulation is excessive and discriminates against commercial banks. The information requested of commercial banks by the Superintendency is unnecessarily large, costly to obtain, and almost impossible to process to serve a useful purpose. Prudential regulation is inadequate; it is not actively pursued and does not provide financial transparency or discipline. Universal banking may be a desirable final outcome, but not if it is obtained through the formation of scattered non-legally related financial institutions.

3.11   The country's regulatory framework is set by the Central Bank Law and the Banking Law, both now under revision. The Central Bank promulgates and implements monetary policy; the Superintendency of Banks (a dependency of the Central Bank) supervises and ensures that laws and regulations are being followed.

3.12   The regulatory framework differs substantially among institutions. For example, the minimum capital requirement is now G600 million for commercial and investment banks[30], and an additional G126 million is required for the right to operate in foreign exchange, but only G100 million for finance companies and G200 million for savings and loan associations. At end-1990, reserve requirements were 37 percent on all demand and saving deposits in domestic currency, 15 percent for the new CDs mobilized by banks, and 20 percent for deposits in foreign currency. Requirements were only 5 percent for deposits in savings and loan and finance companies. The interest rate lending ceiling was eliminated in October 1990; deposit rates have been unregulated since early in the year. It would be important to implement similar reserve requirement regulations among institutions with similar roles. The new Banking Law now being drafted might be the right instrument to initiate this process.

3.13   Banking supervision is carried out under two different procedures: off-site supervision and on-site inspection. The first requires reviewing balance sheets provided by banks on a daily, monthly, quarterly, and annual basis. At present this process is carried out manually; but a study is underway to computerize the information input and the operational procedures. A manual containing methodologies, analytical techniques, and operational procedures is already available. Information requested from banks and provided by them is extremely detailed and most of it is not analytically useful, but verifying compliance with detailed Central Bank economic regulations makes it indispensable. Most of this effort is of little practical use because the lack of transparency in the information available renders it almost meaningless (it is not checked by independent external reviewers and may not follow satisfactory accounting standards). On-site supervision, the second procedure, is conducted by inspectors who visit banks periodically, with inspections limited to headquarters. Unfortunately this most important kind of supervision is not actively pursued. No manuals are available for these inspections and the quality of banks' portfolios are not evaluated.

---

[30]   Central Bank Resolution No. 5 (dated October 26, 1989) raised this minimum to G2,000 million beginning in April 30, 1990 and to G3,000 million starting October 30, 1990. The capital requirements in guaranies deteriorated in real terms until this change. Originally it was US$2.5 million; before the changes it was less than one-third that amount. Capital requirements for finance companies were not changed by Resolution No. 5.

Table 3.1: PARAGUAY - SIZE OF FINANCIAL INTERMEDIARIES
(unconsolidated, as a share of total domestic resource mobilization)a/

	1975	1981	1985	1988	1989	1990
National Development Bank (BNF)						
Liquid Assets	8.7	5.4	5.4	10.2	5.6	5.5
Credit	52.9	22.6	22.7	20.9	21.8	22.1
Dom. Resource Mobilization	14.8	7.8	7.6	10.2	6.3	7.4
Other Net	46.8	20.2	20.5	20.9	21.1	20.3
Commercial Banks						
Liquid Assets	43.2	35.4	45.6	36.5	36.8	30.7
Credit	70.3	72.3	57.6	60.4	68.0	79.2
Dom. Resource Mobilization	73.8	68.7	71.9	71.8	75.6	76.7
Other Net	39.6	39.0	31.2	25.1	29.2	33.3
Savings and Loan Associations						
Liquid Assets	3.3	4.6	2.4	2.2	2.2	1.9
Credit	8.1	16.1	11.3	10.9	11.0	11.1
Dom. Resource Mobilization	10.8	21.1	16.9	14.4	14.0	13.8
Other Net	0.6	-0.4	-3.2	-1.2	-0.8	-0.8
Private Development Banks						
Liquid Assets	0.2	0.0	0.4	0.7	0.6	0.8
Credit	3.6	0.8	0.8	1.1	1.1	1.8
Dom. Resource Mobilization	0.3	0.0	0.0	0.1	0.3	0.6
Other Net	3.6	0.8	1.2	1.7	1.4	2.1
Finance Companies						
Liquid Assets	0.0	0.4	0.4	0.5	0.6	1.1
Credit	0.5	7.6	6.8	5.9	5.6	6.2
Dom. Resource Mobilization	0.3	2.4	3.6	3.6	3.8	1.6
Other Net	0.2	5.6	3.6	2.8	2.4	6.3
Total						
Liquid Assets	55.4	45.7	54.1	50.0	45.7	40.1
Credit	135.4	119.5	99.1	99.3	107.4	121.1
Dom. Resource Mobilization	100.0	100.0	100.0	100.0	100.0	100.0
Other Net	90.8	65.2	53.2	49.3	53.2	61.2

Source: BCP, 1990

a/ Note: Liquid assets plus credit equal domestic resource mobilization plus net other.

3.14   Concerning risk classification of assets, there is a strongly enforced regulation requiring that loans more than 30 days overdue cannot accrue interest, and forcing banks to take legal action after a specified time. However, banks can easily cover-up bad loans by continually rolling them over or by arbitrarily reclassifying them as good, thus avoiding further inspection. Moreover, the Superintendency does not require that a provision be made for nonperforming assets, and there are no effective lending limits to shareholder-related interests. Many banks lack the capacity to properly monitor the quality of their portfolio; one-third of the banks do not use external auditors.

3.15   Due to economic overregulation, lack of mechanization and insufficient personnel, supervision has not been carried out properly. It is estimated that in 1989 only 30 percent of the Superintendency's responsibilities have been performed. This is reflected in inefficiencies in supervisory practices, with approval of banks' balance sheets and thus distribution of dividends to shareholders also affected. It would be desirable to rationalize and simplify Central Bank economic regulations regarding commercial banks. If the Central Bank regulates only the most basic elements dealing with monetary control, the system would be able to concentrate on supervising the intermediaries' financial health and deposit security. This would cut the Superintendency's human and computational requirements would be cut while contributing more efficiently to overall development.

3.16   **Health of the Banking System.** Banks did not correctly anticipate and were strongly affected by the collapse of the fixed foreign exchange rate in 1982, after decades of a constant exchange rate. Although in the years prior to 1982 only a few institutions captured savings in foreign currency and lent the proceeds in domestic currency, most did not take foreign exchange risks directly but lent in the same currency in which they received the funds. Despite this, the crisis was widespread and almost everyone was affected. Debtors not directly linked to foreign trade were affected most since relative prices moved strongly against them. Importers also were affected because of price controls that kept the domestic prices of imports from rising as fast as the new free foreign exchange. Even exporters were affected because they were required to return part of their export proceeds to the Central Bank at the low official foreign exchange (implicitly, a substantial tax was imposed on exports). However, now, seven years after the crisis, most banks seem to make good profits and some are expanding very fast; thus, the crisis of 1982/83 seems over.

3.17   Nonperforming loans rose massively in 1982/83 as a result of the foreign exchange crisis. Their value almost tripled between 1981 and 1983 rising from 7.6 percent to 19.9 percent of the banks' portfolio (see Table 3.2). The renegotiation of loans, the write-off of others, the capitalization of banks, transfers from the Central Bank through subsidized rediscount interest rates, but, above all, the rise in nominal lending, permitted a consistent reduction in that percentage of nonperforming loans in the ensuing years. This percentage now stands at a value almost as low as in 1980, though because of lack of effective on-site supervision and lack of transparency in official information it is possible that the data shown may underestimate the actual size of bad loans.

Table 3.2: **PARAGUAY - NONPERFORMING PORTFOLIO OF COMMERCIAL BANKS, 1981-88**

	1980	1981	1982	1983	1984	1985	1986	1987	1988	1989
A. Billions of Guaranies										
Non-Performing Loans	1.9	6.6	12.7	19.2	19.4	18.2	15.3	12.9	11.4	17.3
Total Portfolio	86.0	107.5	114.1	129.0	151.6	165.4	210.8	254.7	325.5	608.8
Loan Loss Provision	0.4	1.0	1.4	2.3	3.0	2.8	2.7	3.0	2.5	4.3
B. Percentages										
Non-perf/Total (%)	2.2	6.1	11.1	14.9	12.8	11.0	7.3	5.1	3.5	2.8
Less Prov/Non-perf.	21.1	15.2	10.9	12.2	15.4	15.6	17.8	23.2	21.6	25.2

Source: Central Bank and World Bank estimates.

3.18   On the surface, the system as a whole seems financially sound. The bulk of commercial banks appear to be profitable and capitalized. This reflects in part the relative stability of this economy over an extended number of years. The banking sector does, however, contain several very weak institutions. Some private banks are clearly bankrupt and the official banks have lost substantial amounts of resources. There are three private institutions that have negative net worth, but others may also be in trouble. For example, there are two other institutions where "other assets accounts" and unsettled "pending balances to reconcile between branches" (where bank losses may be hidden) relative to equity have more than twice the importance these accounts have on the average of all banks. On top of these, there are four other banks with negative operating margins[31] (except for foreign exchange profits that are not checked by the Superintendency). These nine banks offer about one third of private bank's total credit[32] (only three are foreign owned).

3.19   Until June 1990, regulations specified that capital (and reserves) had to be at least 15 percent of total assets, excluding reserves in the Central Bank.[33] Most banks did not comply with this requirement, which was unrealistically high by international standards. As of the end of June 1990, at least 15 out of 23 banks were in noncompliance; with a capital shortfall of about G30 billion. The capital/asset ratio for the consolidated banking system was just 11.3 percent.

---

[31]   The three banks openly in trouble also show very low or negative net operating margins. The two other banks with large "uncertain" assets show healthy profit margins before taxes but only because of large foreign exchange profits (without these profits, one shows losses, the other has a modestly positive margin).

[32]   Of course, all these institutions need not be corrupt. As explained below, the four with negative profit margins have risk-adjusted capital/asset ratios above 10 percent and of the two with large uncertain assets, one has a ratio above 10 percent. Of the three bankrupt institutions, one if foreign-owned and should not have difficulty arranging recapitalization.

[33]   Assets are defined excluding reserve requirements and cash in foreign and domestic currency. Capital does not include accumulated profits nor does it reduce accumulated losses; the latter exceeded the former between 1983-87, but the situation reversed in 1988. Official regulations deduct the losses but do not allow retained profits to be added to capital.

However, no penalties were applied; instead, the Central Bank reduced requirements by allowing the deduction of rediscounts from the assets determining capital requirements. This is policy probably gives an incorrect signal--not only should capital requirements be enforced, but banks are responsible for repayment of both their normal and rediscounted portfolio and thus should maintain capital against both. There are other changes to legal requirements that could be worth implementing: first, the inclusion of retained earnings as capital; and second, allowing inflation-related revaluation of fixed assets in the definition of capital (and assets) without tax penalties. Assets also could be weighted by risk, to give a better idea of the portfolio's quality. These changes would provide a better picture of the true financial condition of banks. Rough estimates of capital in relation to a risk adjusted portfolio are close to 15 percent, which is the legal requirement. Excluding the nine banks mentioned above, only two show a capital ratio below 10 percent (both foreign owned). Of the nine banks in possible trouble, the three openly so show of course a negative capital coefficient; of the two with large uncertain assets only one has a very low coefficient; the four other banks with negative profit margins all have coefficients higher than 10 percent.

3.20   Accounting practices are poor and some regulations have allowed banks to show unrealistic results. Banks could use different foreign exchange rates to value their foreign currency accounts and some apparently did. With the multiple exchange rate system in place until 1989, banks could choose among many different official exchange rates to value assets and liabilities and the Superintendency could not check this, since it only requested information in local currency. Now the Superintendency has forced banks to value all their foreign currency operations at prevailing market exchange rates. But because it does not require that this information also be provided in the original foreign currency denominations, it is still impossible to check the veracity of the figures provided. The analysis of this latter information seems critically important, as so many banks show large foreign exchange profits that have not been checked.

3.21   **The Central Bank and Banking Laws**. The country now has an excellent opportunity to implement changes in the financial sector through the right drafting of these two laws and their complimentary legal provisions. With regard to bank supervision and the role of Superintendency of Banks (SB), the legal framework should: (i) guarantee the SB's professional independence from the Central Bank; (ii) emphasize the SB's role in enforcing compliance with prudential (not economic) regulations; (iii) encourage better focused on-site bank supervision; (iv) promote use of internationally accepted accounting practices and risk exposure guidelines, (v) define prudent limits on credit granted to conglomerates and shareholders (and their related interests); (vi) define capital requirements and enforce compliance with them, (vii) define a clear-cut calendar of actions to deal with troubled financial institutions and give BS enough power to implement them when these cases arise; (viii) set stringent conditions before the Central Bank can offer financial support to troubled institutions; and (ix) apply homogeneous regulations to similar institutions and allow universal(as opposed to specialized) financial institutions to develop as warranted by the market.

3.22   **Official Banks**. As mentioned earlier, the official banks are the Banco Nacional de Fomento (BNF), Fondo Ganadero (FG), Banco Nacional de Ahorro y Prestamo para Vivienda (BNV), and Banco Nacional de Trabajadores (BNT). They have generated large losses owing to low interest rates (relative to the cost of funds), high administrative costs and poor collection rates. Moreover, these banks also have received implicit subsidies through government recapitalization at no cost. Recapitalizing these banks implies a large commitment; hence it is worth reexamining their roles at this point.

3.23   The BNF has been capitalized several times and recently received another injection of resources from the Government (this time expected to be provided in periodic installments). Its profitability has been estimated to be highly negative. The main problems have been large arrears on its lending, heavy administrative costs, and dependency on foreign resources to be channeled to domestic markets at fixed interest rates in domestic currency. The BNF did not lose even larger amounts because of the implicit subsidy it received, and continues to receive in reduced amounts, from the Central Bank through rediscounts.

3.24   The BNF has been expected to provide medium-term financing for productive sectors and formal credit to small farmers. It has not been successful as a medium-term financing institution, but it has reached the small farmer. However, this has been an expensive endeavor. If the bank can be made profitable, while simultaneously reducing its dependency on foreign financing and increasing its ability to mobilize domestic savings channeled as medium-term loans, the BNF could become a key pillar in the country's development and deserves the strongest support. However, if reforms do not succeed in reaching these goals and the bank keeps losing vast amounts of money, alternative solutions should be sought--for example greater reliance on private banks with limited, targeted subsidies--to improve collections and keep subsidies clear and transparent.

3.25   FG's role has been to offer medium-term credit and technical assistance to large ranchers. It has been successful at this task, but at a heavy financial cost. FG has almost exclusively relied on foreign funds and, like the BNF, has often lent at fixed interest rates in local currency. As a result, foreign exchange loses have been substantial. If its foreign financing is valued at market exchange rates (some loans have been undertaken by the Finance Ministry, which formally carries the foreign exchange risk), FG's net worth is substantially negative. Its low profitability suggests poor prospects.

3.26   The FG has played a valuable role in the country's livestock sector, but in the process it has transferred large amounts of wealth from taxpayers to large ranchers. The latter can be avoided in the future if FG's function are handed over to the private sector. If the private sector is not interested in taking such a role, creating a new public (or mixed) bank to carry FG's role is an option that should be strongly discouraged.

3.27   BNV's credit operations should be subject to a low, strict ceiling while its future role is reviewed. At present, it is supposed to promote housing construction while serving as a central bank for the savings and loan institutions. But, because of the country's relatively high inflation in recent years, the lack of inflation-adjusted interest rates, and the lack of the medium-term financing, housing finance is nonexistent and savings and loan institutions operate like other financial institutions controlled by the Central Bank and could be supervised by it. In these circumstances, there is little justification for the BNV.

3.28   BNT operates as a normal commercial bank (except that it used to generate large losses, like the other official intermediaries) but is financed through a tax on wages. There is little reason for an institution like this. The Government should be encouraged to eliminate the tax that finances BNT, reduce BNT's size, and if it cannot be made profitable, close it in a reasonable time-frame.

B. Trade Policy Reform

3.29   Trade related distortions are not important in Paraguay now, but they could become sizable. In practice and in relation to other countries, nontariff barriers are few.[34] However, the existing Customs Law and some related taxes have a heavy protectionist bias. Custom duties (including surcharges) range from 3 percent to 86 percent, and tend to be higher for traditional manufacturing sectors. These duties are consistent with high effective protection rates. In fact, it would not be surprising to find a few subsectors operating with a negative domestic value-added if their output and inputs were valued at international prices.[35]

3.30   However, because the Government operated under simple, special tariff regimes and because smuggling was widespread, the Customs Law was not effective. Despite the notionally high tariffs, ordinary duties (those defined by the Customs Law) represent just 7 percent of taxable imports [36] (registered imports were close to 90 percent of estimated imports in 1990, but were a much lower percentage in earlier years). Thus, indirect evidence implies that actual protection is much lower than the law suggests. To a significant extent, Paraguay seems to be <u>de facto</u> a free-trade economy with low actual tariffs and taxes.

The Present Customs Law

3.31   Although distortions induced by the current, effective trade regime may not be large and the law does not reflect actual outcomes, it is still important to study the implications of properly applying the law as stated. This is of special relevance now because the present political system is evolving towards a full democracy, and economic policies may change as new authorities are elected at different levels. If the existing legislation is potentially disruptive, then it is urgent to change it to a tariff code that more closely reflects current reality; any incoming administration might try to enforce it, thus causing costly and long-lasting negative effects on growth, employment creation, and income distribution. Moreover, moving toward lower, more unified tariffs would be consistent with the changes taking place in Argentina, Brazil and Uruguay and thus leave Paraguay in a better position for its entry into the MERCOSUR.

---

[34]   Some basic foods are subject to trade restrictions at the harvesting season, but they seem to have little practical relevance. Many imports were prohibited before 1989 and smuggling was widespread; at present, some import restrictions remain, initially set in Decree 1663 of December 1988, Article 9, and as before, the corresponding goods are widely smuggled. Despite this, the policy implemented in 1989 is far more efficient.

[35]   These are activities that operate with a high content of imported inputs and whose value at international prices is higher than the import value of the final product (e.g., assembly industries for bicycles).

[36]   Total taxes on imports (including taxes on imports collected by Customs but labeled as internal taxes rather than duties (see para. 3.33)) represented 12 percent of registered imports in 1989 (see Table 3.4).

3.32   In December 1988, Paraguay adopted the Harmonized System for tariff classification, which is more complete but also more complex than the previous one. The old system had 14 different tariff levels while the new one had 41 as of December 1989. The new system was hastily introduced, so Customs personnel and related staff still lack training to use it fully. In many cases its application is too discretionary and thus often inconsistent.

3.33   Fortunately, the application of the ordinary customs regime of duties often is disregarded and replaced by special regimes. About 25 percent of imports from neighboring countries (affecting mainly Argentina and Brazil) pay a flat 10 percent tax; items associated with "tourism" pay 7 percent following transitory regulations that are not part of the Customs Law. (Imports of scotch and cigarettes--two tourism-related items--are subject to another special procedure.) Only half of the 10 percent tax on imports from neighboring countries is considered an import duty; the other half is labeled "internal tax" because this import tax was enacted to replace certain internal levies that were eliminated simultaneously. Regarding the tourist regime, five points of the seven are considered import duties and the other two internal taxes. In Table 3.3, the five percentage point taxes on imports from neighboring countries and tourist goods are shown as part of ordinary duties, the remainder as "internal taxes"; in Table 3.4, the "internal taxes" are included in the total import tax rate, but not the ordinary rate.

3.34   There also are additional taxes on imports <u>that are not collected by customs</u> and, correspondingly, often evaded. First there is a surcharge of 3 percent on raw materials and capital goods, of 30 percent on beverages and tobacco, and of 6 percent on goods not under the 3 percent or 30 percent rates. Since this surcharge on imports is considered an "internal tax," it is <u>not</u> charged at customs and often evaded. In addition, while domestic output is nominally taxed at 4 percent under a value-added type of sales tax (which is almost always evaded), imports are taxed at 8 percent and 14 percent. The higher rate (another tax on imports <u>not</u> collected by customs) is considered an internal tax and also is evaded, though less frequently than on domestic goods. About 80 percent of the revenue from the sales tax come from imports and thus the sales tax is mainly an import tax.

Revenue Collection

3.35   The Authorities have succeeded in making important changes in custom duties to increase revenues and reduce corruption and evasion. Among others, starting in May 1989, taxes on imports were applied to import values at the free exchange rate.[37] As a result, collections of import taxes more than doubled in 1989. (This figure includes the taxes on imports collected by customs that are legally labeled "internal taxes" (see para. 3.33 and Table 3.3). Total taxes on imports rose to 22 percent of (a much higher) tax revenue in 1989, compared to 17-18 percent in 1986-88. Import duties were 25 percent of total taxes after May 1989, when the new valuation system started operating in full. This percentage would be even higher if other duties applied at sales time only to imported products, were classified under this latter category.

---

[37]   For customs purposes, the value of the US$ was G400 in 1988; G550 in January and February 1989, G750 in March and until May 9; from May 10 on it has been the free exchange rate (around G1,200 in December 1989).

3.36   Two factors helped increase import tax revenues: first, the taxable value of imports increased because of the higher value of foreign exchange used; second, the Authorities simplified the tax system and reduced tax rates. Average tariff rates according to the tariff codes were reduced around 25-30 percent; the average rate paid for ordinary import duties declined from 10 percent to 7 percent, and the overall rate of taxes on imports, including "internal taxes" went down from 16 percent to 12 percent (Table 3.4).

3.37   Efforts to simplify the system included widening the application of a 10 percent tariff on selected goods from neighboring countries (Brazil and Argentina). Originally, only necessities were eligible for the ten percent rate, but the list has gradually expanded and now covers about a fourth of total trade with those countries. The base for the tourist regime (a 7 percent flat rate) was also widened. It was initially applied to products linked to tourism in border cities but is now in place throughout the country. It affects tourists and locals and covers a wide variety of products including some unlikely to be related to tourism (such as odontological equipment).

Table 3.3: PARAGUAY - CUSTOMS COLLECTED TAXES ON IMPORTS
(in Billions of Guaranies)

	1985	1986	1987	1988	1989
Total Taxes on Imports	15.2	21.6	30.9	42.4	89.6
Ordinary Duties a/	11.1	15.5	19.2	26.0	53.3
Special Revenues b/	0.8	0.9	1.5	2.3	1.8
"Internal taxes" c/	3.3	5.2	10.2	14.1	34.5
Total Taxes	96.3	125.0	177.9	233.0	400.8
Import Taxes/Total Taxes(%)	15.8	17.3	17.4	18.2	22.4

Source: Finance Ministry, General Directorate of Customs.

a/ Includes normal customs duties and the "customs" fraction of the neighboring countries' trade and tourism regimes.
b/ Special regimes of minor importance.
c/ Includes import taxes labeled "internal," namely the part of the taxes on tourist goods and trade with neighboring countries that replaced internal taxes, the stamp tax applied to imports, and other special import taxes.

3.38   The evolution of tax revenues from these two special import taxes has been remarkable (see Table 3.4). While ordinary import taxes doubled, because the base for special import taxes was widened, collection from the two special tax regimes probably increased much more (Data on taxes on imports are not discriminated by regime in 1988). Although the value of taxable imports subject to normal duties almost doubled,[38] border-trade imports increased 250 percent and tourism imports increased almost tenfold. The number of statements from persons and entities using the "neighboring-countries" provision also increased tenfold between 1988 and October 1989.

3.39   In 1989, one-third of ordinary customs revenues came from these two special regimes but this is an underestimate. If the portion of these taxes labeled "internal" also were classified as import taxes,[39] these two special regimes would account for close to 6 percent of total taxes.

3.40   The lower special tax on neighboring countries' imports may be counterproductive for Paraguay by diverting legal and/or illegal imports from less expensive sources in third countries to legal but more expensive sources from Brazil and Argentina. In this case, the Government would receive less revenues and/or the "smuggling premium" (a source of income for some nationals and a potential revenue source for the Government) would decline as the activity becomes legal. Unfortunately, it is not yet possible to evaluate the dimension of the trade diversion effect.

---

[38] Of course, this includes the effect of the reduction in non-registered imports resulting from a unified exchange rate and simpler regulations, which reduced incentives to smuggling.

[39] These taxes are included under the label "Non-Custom Duties" in Table 3.3.

Table 3.4: PARAGUAY - ORDINARY IMPORT DUTIES
(in Billions of Guaranies and percent)

	1988	1989
	Guaranies	
Taxable Imports	<u>264.4</u>	<u>737.9</u>
Normal	220.3	412.2
Tourism	25.4	261.1
Border trade	18.7	64.6
Tax Collection	<u>26.0</u>	53.3
Normal	NA	35.3
Tourism	NA	14.2
Border trade	NA	3.8
	Percent	
Average Ordinary Rate	<u>9.8</u>	<u>7.2</u>
Normal	NA	8.6
Tourism	NA	5.4
Border trade	NA	5.9
Memo Item		
<u>Total Import Tax Rate</u> a/	16.0	12.1

Source: Ministry of Finance, Directorate of Customs.

a/ Total import taxes include taxes that conceptually
(though not legally) belong in this category. See Table 3.3.

Intended Effect of the Customs Law

3.41   The high tariffs in the customs code serve little purpose. The code has many high tariffs,[40] so that the average of all the ordinary tariff codes is 16.2 percent (see Table 3.5), while actual collection of import duties is only 7.2 (see Table 3.4). This discrepancy, which occurs in most countries, suggests that the code's high tariffs are not there to increase revenues, but intended to provide protection. This finding is confirmed by the weighted average tariff calculated using actual imports (December 1989) as weights. This average is 8.3 percent, which is

---

[40] The maximum rate is 72 percent, which grows to 86 percent if surcharges are included.

higher than the observed value but consistent with it because only about 80 percent of potential revenues are collected, the rest being legally exempt.

Impact of Customs Duties by Sector

3.42   For agriculture, the average tariff on all related codes is 18 percent but the import-weighted average is only 8 percent. A similar discrepancy takes place for manufacturing, where the simple average is 16 percent while the weighted one is 8 percent. Mining on the other hand has small tariffs on most codes and thus shows little discrepancy between the two averages (see Table 3.5).

Table 3.5: PARAGUAY - IMPORT TARIFF RATES BY SECTOR
(Weighted by Import Values in December 1989)

Goods	Simple Average Tariff	Weighted Average Tariff a/	Imports Share b/
Overall Average	16.2	8.3	100.0
Agriculture	18.2	7.8	0.3
Mining	4.2	3.6	0.3
Manufacturing	16.3	8.3	99.4
Manufacturing by Use	16.2	8.3	100.0
Consumer Goods	23.9	8.7	54.1
Intermediate Goods	11.5	5.6	17.3
Capital Goods & Transport Equipment	12.6	9.1	28.6

Source: World Bank calculations using customs data for December 1989.

a/ Using tariff rates adjusted by special regimes (tourism, 7 percent; border trade (10 percent).
b/ Manufacturing disaggregates add up to 100 percent.

3.43   In manufacturing, few activities benefit from high tariffs in the customs code. As a landlocked country facing high transport costs, Paraguay naturally has its domestic activities protected from external competition. But this characteristic also discriminates against exports.

3.44   Using the three-digit International Standard Industrial Classification, less than 0.5 percent of total imports come from subsectors with average tariffs over 30 percent, and 2.5 percent of imports come from subsectors with average tariffs above 15 percent. The only items benefiting from tariffs over 15 percent are soft drinks, motorcycles, textiles, tanneries, and some wood (furniture), paper, and clay products.

3.45   Soft drinks may not always be legally brought into the country, but are abundantly imported through informal channels. The domestic industry is already able to compete with those imports and a reduction of tariffs would have little effect on domestic prices or production. Motorcycles, like cars, are also widely brought into the country, though not through Customs. Duties on cars were recently reduced, but it remains to be seen if registered imports of cars will increase. There is no domestic production of these goods, so the Government is the only one losing from illegal imports.

3.46   Some textile activities are protected, with rates reaching close to 30 percent. However, textiles also are often imported through informal channels and domestic production has dealt with this challenge. In fact, some enterprises are already exporting textiles, which is not surprising, given the country's comparative advantage in producing cotton; so it is unlikely that domestic output will suffer from a substantial tariff reduction. Textile activities are expanding despite the present system of temporary imports that discriminates against them: exporters of apparel can deduct taxes paid by foreign producers (import taxes on inputs are exempt) but cannot deduct taxes paid by local textile manufacturers. As the apparel industry is growing fast, local textile industries would be better off with lower tariffs on their final products if this were also accompanied by less discrimination against their output through existing export incentives.

3.47   Tanneries also seem to receive rather high protection rates, while leather products and footwear receive very little. If this is correct, it is not only inefficient, but punishes industries that are more important at the national level (e.g., the shoe industry alone has a national value-added larger than the other two combined). Thus, the economy would benefit if protection of tanneries is substantially cut and brought in line with that of shoes and leather products.

Table 3.6: PARAGUAY - WEIGHTED AVERAGE TARIFF RATES FOR MANUFACTURING

CIU Classification	Simple Average Tariff	Import Weighted Average		Production Weighted		
		Tariff a/	Weights	Ordinary Tariffs	Including Spec. Regimens a/	Weights
Food, Beverages and Tobacco	22.0	6.1	12.6	25.8	7.3	49.9
Textiles, Apparel, and Leather	25.6	13.8	5.2	31.0	14.1	11.1
Wood and Furniture	37.0	33.7	0.0	29.8	15.5	13.6
Paper, Paper Prod. & Printing	18.2	5.3	2.0	31.9	11.5	3.6
Chem., Coal, Rubber & Plastic	6.5	5.9	17.8	6.1	4.6	11.2
Non-Metallic Minerals	19.8	11.2	1.6	22.9	13.8	4.6
Basic Metal Industries	8.4	4.4	2.7	7.8	3.3	0.5
Metallic Ind., Machinery & Equipment	14.6	9.5	50.5	16.7	11.9	4.7
Other Manufacturing	21.8	7.1	7.6	24.2	7.1	0.8
Average	16.2	8.3	100.0	23.4	9.5	100.0

Source: World Bank calculations based on Customs data.

a/ Using tariff rates adjusted by special regimes (tourism, 7 percent; border trade, 10 percent).

3.48   Wood furniture and clay products are other activities that might benefit from high tariffs. Further study is needed to be able to say if they will be significantly affected by a drop in tariffs. With regard to the paper industry, it has segments that are also competitive with production from abroad, it is beginning to export. For example, printed books are being exported to other countries in South America; notebooks compete favorably with inexpensive products from Brazil and Argentina. This industry has had substantial protection of little relevance, since smuggling from neighboring countries is almost impossible to stop. It is likely that reducing tariffs in these industries will have little negative effects; instead, the economy will benefit from this change by speeding the transfer of resources into activities where the country has more comparative advantage.

3.49   The small incidence of high tariffs on domestic output is corroborated when examining more aggregate industrial data. Table 3.6 illustrates average tariffs by main manufacturing subsectors. In it, one can again see that simple averages of ordinary tariff codes are much higher than the import-weighted averages (except for chemicals), with tariffs in two subsectors exceeding 25 percent. In the import-weighted averages, except for wood and furniture, no aggregate exceeds 15 percent and most are below 10 percent.

3.50   The disaggregate figures in Table 3.6 also support the view that the ordinary customs schedule tends to be much more protectionist in theory than in practice. The average production-weighted tariff in the ordinary schedule is 23.4 percent, with most activities showing tariffs higher than 25 percent. By contrast, the tariff schedule actually applied shows little protectionist tilt. Because of the special regimes on border trade and tourism, the production-weighted average is only 9.5 percent, just slightly higher than the import-weighted average previously mentioned (8.3 percent). Moreover, the production- and import-weighted averages give low and similar results for the subsectors. It is worth noting that the production weighted-tariff in the wood and furniture subsector is much lower than the import weighted tariff, since in this activity high tariffs protect enterprises with minor domestic output.

Informal Imports

3.51   Informal activities (smuggling among them) are important in Paraguay. Leaving aside their negative impact on the legal system and the ethical question, illegal imports have had some positive indirect economic effects. Contrary to the experience of most Latin American countries, production inefficiency has had a ceiling set by apparently low smuggling costs (estimated at about 10-15 percent of CIF value).

3.52   Off-custom imports are an element of several informal activities carried out not only by small enterprises but also by many large ones that operate using a parallel financial accounting system. Parallel accounting is the enterprises' reaction to what they perceive as a heavy burden implicit in most tax laws. Managers of private firms argue that income tax rates on profits are too high (30 percent), that some indirect taxes are also too high (for example, 30 percent on some imports), and that the burden of taxation is not evenly distributed among the productive sectors. Agriculture is especially privileged in the sense that it is exempt from most taxes and receives a large share of subsidized credit.

3.53   The private sector also argues that informal imports would not stop even if all import tariffs were abolished, as long as other taxes remain high. A reform of trade taxes may not produce substantial revenue unless this action is accompanied by a general tax reform the private sector considers fair. In fixing tax rates the government must therefore be cognizant of its ability to collect taxes and not set rates that will encourage evasion, at the cost of public revenues.

The Challenges of MERCOSUR

3.54   The recently signed MERCOSUR treaty between Brazil, Argentina, Uruguay and Paraguay reducing duties on imports from third countries and eliminating tariffs among signatory countries will pose special challenges to Paraguay. One possible result of the pact might be high external tariffs--this would hurt Paraguay if these tariffs were applied, but current experience suggests this is unlikely. However, if the external tariff is low, then Paraguay may be forced to make adjustments in informal activities and government revenues that are much more significant than for the other countries.

3.55   Paraguay has profited from high protection in Brazil and Argentina and its informal activities have flourished by taking advantage of such conditions. If protection in those countries drops substantially, then Brazilian and Argentinean customers will not need the services now informally provided by trade in Paraguay, and these activities will suffer. In addition, duties stemming from legal imports from Brazil and Argentina are of significance for Paraguay and these duties will be eliminated when the treaty is implemented. Paraguay needs to carefully prepare its economy to absorb the impact of MERCOSUR. One way would be to move toward lower, more uniform tariffs, which would not only reflect reality, but make Paraguay's tariffs more consistent with its MERCOSUR partners. The Government also will need to find alternative forms of revenue. More than 6 percent of the country's taxes stem from duties on Argentinean and Brazilian imports and goods imported for sales to tourists that will no longer come to Paraguay. New productive activities will need to be identified and expanded to absorb the resources left unemployed by the contraction of informal activities. If subsidies are to be avoided in reaching this latter goal, studies should be advanced and the corresponding information generated widely shared with the private sector.

Export Incentive Policies

3.56   Paraguay uses a special drawback scheme to benefit exports. Based on a system of "temporary imports" it essentially refunds import taxes. Though the laws regulating the system have been in place for some time, its acceptance has increased strongly only since 1988. The new Investment Law (see next paragraph) further encourages its use. However, temporary-import policy discriminates against domestic inputs and stimulates vertical integration because the Government returns to exporters the import taxes that they pay but not the taxes they pay on inputs bought from local producers. These latter taxes can only be avoided by backward integration of the activity. This may help explain why Paraguay exports raw cotton and soybeans but has failed to develop a sizable export industry of finished products. Producers in the garment industry complain about this problem; it affects the profitability of their business and works against export diversification.

3.57   An Investment Incentives Law was enacted in 1989. It gives beneficiaries 5-year tax holidays on 95 percent of income taxes and 6 months of duty-free imports; however, the law's regulations have not yet been issued. The system seems discretionary and poses serious problems that can be translated into significant tax revenue losses, resource misallocation (creation of artificial new firms instead of expanding existing ones), and incentives to close existing firms at the end of the tax holiday. A group of privileged entrepreneurs may emerge with a constituency for perpetuating interventionist policies.

Summary

3.58   In practice, Paraguay operates with tariffs that are low and quite homogeneous, despite a Customs Law often suggesting high tariffs (30 percent and sometimes reaching over 70 percent) with wide dispersion. Three factors have helped achieve low, homogeneous tariffs in practice. First, simple special regimes operating with low and flat rates (10 percent for some border trade; 7 percent for "tourism" trade) have replaced many ordinary tariffs (often the high rates). Second, taxes often reaching close to 5 percent of imports are charged under different names even on tariff-free items; therefore, even though the lowest tariff is often zero, some taxes are paid even in these cases. And third, unregistered imports set a ceiling on tariff rates. If tariffs exceed 10-15 percent, goods tend to be imported through informal channels.

3.59   In light of the regime of effective tariff rates, as opposed to the tariff code, it would be simpler for Paraguay to operate with one flat across-the-board tariff rate of, say, 10 percent, replacing all present import taxes. This policy would not preclude further consumption taxes on specific items (e.g., luxuries) that can also be charged at customs and on domestic producers, and "without" exemptions.[41] Since imports represent about 30 percent of GDP, import tax revenues would be about 3 percent of GDP--somewhat higher than now collected (2.4 percent)--and the system would be more efficient and less prone to corruption.

3.60   Few sectors would be affected by these a changes, since the economy is already operating effectively under a similar regime in practice. However, some enterprises may suffer. In the few cases where lower tariffs will do significant damage, an accelerated schedule of tariff cuts can be agreed with the producer until the common flat rate is reached. However, for lower import taxes to be successfully collected, a far-reaching general tax reform will need to be accompanied by elimination of discriminatory tax-incentive schemes (e.g., the Investment Law). The recently signed MERCOSUR treaty with Argentina, Brazil and Uruguay poses more difficult challenges. Government revenues may suffer and informal activities will be curtailed if the treaty is successfully implemented. Paraguay will need to study how to best take advantage of the treaty's favorable provisions and compensate for the short-term adjustment costs.

C. Public Enterprise Reform

3.61   Paraguay has thirteen large public enterprises: four in the transport sector (two airlines, LAP and LATN; railroads, FCCAL; and riverways, FLOMERES); one in the petroleum refining and distribution area (oil and gas, PETROPAR); three in the manufacturing sector (steel,

---

[41] It will be impossible to avoid some exemptions because of international agreements, but they can be minimized.

ACEPAR; cement, INC; and alcohol, APAL); two administrations (airports, NAC; and ports, ANNP); and three utilities (electricity, ANDE; water and sewage, CORPOSANA; and telecommunications, ANTELCO). A holding company (SIDEPAR) has just been divested.

3.62   The financial condition of public enterprises is weak despite some improvement in 1990. Public enterprises still show a deficit well above that observed at the beginning of the 1980s, and as with the government, their external debt is not being serviced on time. Public enterprise finances received a severe blow in 1989 with the exchange rate liberalization in February. Some enterprises had been heavily subsidized in their use of foreign exchange to buy imports and service external debt through the now-abandoned system of multiple exchange rates. In addition, there were substantial wage increases in the first semester of 1989, and adjustments in tariffs were well below initial programs. As a result most enterprises' financial performance deteriorated in 1989 but appears to have improved in 1990 (see Table 3.7) as a result of lower investments in some enterprises, improvements in efficiency, and higher tariffs[42]. However, the current financial performance remains unsustainable, and much remains to be done to reach satisfactory results.

3.63   Public enterprise financial performance generally deteriorated during the 1980s despite the foreign exchange subsidies, mainly because of large increases in capital expenditures-- some profitable (electricity and, to a degree, telecommunications), others with much less return (river transportation), and still others in overdimensioned and obviously unprofitable endeavors (steel and cement). The larger capital outlays were financed mainly overseas. The enterprises encountered problems in servicing this external debt and accumulated arrears even before 1989; after 1989 it became even more expensive to service the external debt because of the higher real exchange rate.[43] Resolving the problems posed by this debt and the unproductive and overdimensioned investments financed abroad, will involve actions by Paraguay and cooperation by the creditors. Brazil started the process and gave Paraguay an excellent deal on its debt which benefitted mainly ACEPAR (see paras. 1.53 and 3.92). A Paris Club agreement would help Flomeres and INC. Nonetheless, such agreements generally are extensions of amortization periods, not debt reductions. The government will need to take action to improve public enterprise performance and thereby reduce the need for additional financing of public enterprises.

---

[42]   Investment is usually underestimated in preliminary reports due to data collecting difficulties in measuring expenditures financed abroad (mainly capital).

[43]   Due to the large voluntary and involuntary foreign financing (i.e., accumulation of arrears), public enterprises did not need to increase their net indebtedness with the domestic banking system in 1989-90. Had public enterprises made their foreign interest and amortization payments on schedule, significant credit expansion from the banking system would have been necessary.

Table 3.7: PARAGUAY - PUBLIC ENTERPRISE FINANCIAL ACCOUNTS
(% of GDP)

	1980-81	1983-84	1986-87	1988	1989	1990 p/
Value Added a/	3.0	3.4	4.1	4.2	3.9	6.5
Current Expend.	1.4	2.1	2.7	2.7	4.1	5.3
Wages	1.2	1.5	1.3	1.5	1.6	2.2
Interest b/	0.2	0.5	0.5	0.7	1.5	1.7
Taxes	0.1	0.1	0.8	0.6	0.9	1.2
Other	0.0	0.0	0.1	0.1	0.1	0.1
Own Current Savings	1.5	1.3	1.4	1.5	-0.2	1.1
Real Investment	2.0	4.3	3.4	3.9	4.9	5.6
Other	-0.2	-0.1	0.6	1.1	0.3	0.0
Deficit b/ c/	0.2	2.9	2.6	3.5	5.4	4.4

Source: Statistical Annex, Tables 5.5, 5.6 and 2.1.

a/ Includes other own revenues but is mainly value of sales less cost of intermediate inputs.
b/ On a cash basis until 1988, accrual thereafter.
c/ Revenues from transfers received from the Central Government are not included.
p/ Preliminary estimates.

3.64 The public enterprises' total deficit reached 3.5 percent of GDP in 1988, rose to 5.5 percent in 1989 and is preliminarily estimated at 4.4 percent in 1990 (see Table 3.7). (The figures for 1989 and 1990 refer to an accrual basis, 1988 on a cash basis, but the difference between cash an accrual was not great in 1988 because the low, distorted exchange rate meant that the financial cost of the interest arrears was small.) The public enterprises' current savings are estimated to have declined substantially in 1989 but they rose again in 1990.

3.65 The enterprises with the largest deficits in 1988 were INC, ANTELCO, and ACEPAR. INC and ACEPAR's deficits declined in 1989 and 1990, despite lower savings because their investments also declined drastically with the completion of plants that had been under construction. ANTELCO's performance showed a similar pattern, but the decline in investment reflected cuts in programs underway.

3.66 Despite the improvement in these three enterprises, the savings of the public enterprises fell dramatically in 1989 and the deficit rose sharply, largely due to the results of ANDE and PETROPAR. ANDE started important new works in 1989; thus, despite an increase in savings that allowed it to finance a higher interest bill, it posted by far the highest deficit in 1989 and 1990 (see Table 3.8). PETROPAR's finances deteriorated in 1989 as a resulted of higher taxes paid and insufficient adjustments in fuel prices after the elimination of the multiple exchange system. PETROPAR's large deficit in 1989 (one of the highest among PEs) was also the result of a large increase in inventories. However, the drastic increases in fuel prices that took place in 1990 produced a strong improvement in PETROPAR's savings performance and turned its deficit into a surplus, which explains almost the whole reduction in the deficit of the public enterprises.

3.67   There are some encouraging signs in public enterprise performance. Among them are management improvements and gains in efficiency to a large extent linked to less financial "leakage"[44] and some reductions in future financial requirements reflecting completion of investments at the cement and steel companies--the latter reflected in "other" capital expenditures in Table 3.7 due to accounting problems.

3.68   Despite these signs of improvement, the performance of the public enterprises remains unsustainable. There continue to be major problems with managerial accountability, accounting and control practices, and overstaffing. These problems help explain the poor quality of many enterprise's investment programs and unsound financial results in the 1980s. The Government has began to audit its enterprises but the institutional setup under which they operate is not yet well defined, which makes their control very difficult. Although financial performance appears to have improved in 1989/90, some of this progress may be misleading. Moreover, financial troubles should not be the only concern. Inefficiency is a key issue, which is not fully reflected in the enterprise finances because of the monopoly power that several enterprises hold. Most relevant in this regard are the cases of ANTELCO, CORPOSANA and PETROPAR. The former two can set high enough tariffs to cover inefficient operations (reflected for example, in overstaffing). ANTELCO has already done this, but CORPOSANA's tariffs are too low. PETROPAR might use its monopoly power on gasoline and gasoil to set high enough prices and generate a large enough surplus that would allow it to build a new refinery that probably would not be profitable in a competitive environment; its finances would look good but at a high cost in terms of efficiency. Thus the public enterprise problem is not simply one of raising prices, which would simply shift the burden of their inefficiency from the Government to the consumer, with a corresponding loss of competitiveness.

3.69   Privatization and joint ventures with the private sector (local and/or foreign) are alternatives that should help improve results in several enterprises. A brief summary of the main issues in each enterprise follows.

---

[44]   For example, in PETROPAR expenditures for crude oil purchases did not increase in 1989 while the related exchange rate almost tripled. This implies a radical improvement in management practices. In the Central Government, the Health and Education Ministries are similar examples that are often mentioned.

Table 3.8: MAIN PUBLIC ENTERPRISES FINANCIAL DATA, 1985-88
(% GDP)

	1986/87	1988	1989	1990 p/
Current Own Savings a/	**1.46**	**1.46**	**-0.19**	**1.20**
ANDE	0.62	0.69	0.70	0.69
ANTELCO	0.22	0.27	-0.16	0.36
CORPOSANA	0.04	0.07	0.04	0.08
FLOMERES	0.06	0.04	-0.09	-0.08
INC	-0.06	-0.12	-0.38	-0.38
LAP	-0.09	-0.11	-0.22	-0.44
ACEPAR/SIDEPAR	0.00	0.00	-0.10	-0.05
PETROPAR/APAL	0.64	0.61	-0.08	0.74
Other	0.02	0.02	-0.09	-0.24
Deficit a/	**2.84**	**3.53**	**5.36**	**4.38**
ANDE	0.43	0.48	2.45	3.56
ANTELCO	0.19	0.97	0.55	0.15
CORPOSANA	0.33	0.18	0.21	0.52
FLOMERES	0.00	0.02	0.10	0.09
INC	1.09	1.17	0.57	0.40
LAP	0.37	0.38	0.53	0.55
ACEPAR/SIDEPAR	0.51	0.75	0.23	0.08
PETROPAR/APAL	-0.09	-0.58	0.81	-0.73
Other	0.02	0.16	-0.09	-0.25

Source: Central Bank, Ministry of Finance, and World Bank estimates.

a/ Excludes arrears (in 1986/87 and 1988) and transfers received from the Government.
p/ Preliminary estimates.

ANDE

3.70 The Electricity Company (ANDE) was one of the few public enterprises that performed consistently well throughout the 1980s. It generated current savings of close to 1 percent of GDP, based on a stable value-added and a consistent cost structure (see Table 3.9). However, its deficit multiplied five times in 1989 and grew even larger in 1990 due to increased capital outlays. This reflected a rapid expansion in electricity distribution, the main element being the completion of the Fourth Distribution Line Project linking Itaipu and Asuncion. Nonetheless, some works were deferred due to lack of resources. The preventive repairs on the first hydropower plant in the country (Acaray) are an example. ANDE also has frozen the hiring of new personnel. Investment is expected to drop in 1991 as works are completed. With these lower funding requirements, the deficit should be eliminated and maintenance works might be brought back on schedule.

3.71   ANDE had ambitious investment plans to expand electricity distribution coverage until 1995; however, tight public finances and the strong increase in the exchange rate that ANDE must pay to service its foreign debt has forced the rescheduling of that plan. Works locally financed have been postponed, and those with foreign financing are being spread over a longer period. It is not expected that these difficulties will have long-lasting consequences.

Table 3.9: PARAGUAY - ANDE FINANCIAL ACCOUNTS
(% GDP)

	1980/81	1983/84	1986/87	1988	1989	1990 p/
Value added a/	1.15	1.05	0.93	1.07	1.21	1.47
Current Expend.	0.31	0.34	0.31	0.38	0.51	0.79
Wages	0.25	0.25	0.24	0.30	0.33	0.56
Interest b/	0.05	0.07	0.07	0.07	0.18	0.20
Taxes	0.00	0.00	0.00	0.00	0.00	0.00
Other	0.01	0.02	0.01	0.01	0.01	0.02
Own Savings	0.84	0.71	0.62	0.69	0.70	0.69
Real Investment	0.46	0.87	0.75	0.86	3.15	4.24
Other	0.05	0.11	0.30	0.30	0.00	0.00
Deficit b/ c/	-0.33	0.27	0.43	0.48	2.45	3.56

Source: Central Bank, Ministry of Finance, and World Bank estimates.

a/ Includes other own revenues but it is mainly value of sales less cost of intermediate inputs.
b/ On a cash basis until 1988, accrual afterward.
c/ Revenues from transfers received from the Central Government are not included.
p/ Preliminary estimates.

3.72   As indicated above, ANDE's investment is expected to decline sharply after 1991 as ongoing works are completed; committed new projects with foreign financing (mainly from IDB and OECF) will start more slowly. ANDE is prepared to extend electricity services to interested customers faster than planned if these customers agree to finance the works either directly or through bank loans with ANDE collateral. ANDE would pay the corresponding credit by providing no electricity charges for some agreed time. To meet its external debt, ANDE is paying all interest but no amortization, except to multilaterals and a highly concessionary British loan.

ANTELCO

3.73   ANTELCO, the Paraguayan telecommunications company, experienced a some improvement in its financial accounts in 1990 after a significant deterioration in 1988-89; there is room for still further improvement. Though ANTELCO showed deficits through most of the 1980s, in 1988-89, its deficits were much higher than previously (see Table 3.10). ANTELCO's deficits reflected a substantial investment program concentrated on providing phone service in the Chaco and the eastern regions, especially in 1988. The deficit was reduced somewhat in 1989 by reducing capital outlays. However, in 1990, preliminary data indicate that investment was increased again. With the rise in tariffs and despite an increase in the wage bill and other spending, current savings rose enough to finance most of the higher investment.

3.74   Special care is needed in interpreting the improvement in 1990. Telecommunications is an activity widely regarded as very profitable all over the world--in Paraguay it is a monopoly as well. ANTELCO has the option to increase tariffs to finance current and capital expenditures that may not be economically efficient (overstaffing, fancy equipment). To some extent it has done so in the past. Therefore, it may be desirable to give the Central Government a larger say in the enterprise's investment program, and to set control systems that encourage independence but also accountability. It also is difficult to defend that such a potentially profitable endeavor pays no taxes or distributes any profit to its owner, the Government. Privatization is an option worth studying and some progress appears to be taking place in this regard.

Table 3.10: PARAGUAY - ANTELCO FINANCIAL ACCOUNTS
(% GDP)

	1980/81	1983/84	1986/87	1988	1989	1990 p/
Value added a/	0.92	0.93	0.70	0.73	0.69	1.44
Current Expenditures	0.41	0.53	0.47	0.46	0.85	1.08
Wages	0.34	0.43	0.32	0.33	0.46	0.61
Interest b/	0.07	0.10	0.14	0.12	0.38	0.39
Taxes	0.00	0.00	0.00	0.00	0.00	0.00
Other	0.01	0.01	0.01	0.01	0.01	0.07
Own Savings	0.51	0.40	0.22	0.27	-0.16	0.36
Real Investments	0.49	0.37	0.38	1.22	0.36	0.51
Other	0.01	0.07	0.03	0.02	0.02	0.00
Deficit b/ c/	0.00	0.04	0.19	0.97	0.55	0.15

Source: Central Bank, Ministry of Finance, and World Bank estimates.

a/ Includes other own revenues but it is mainly value of sales less cost of intermediate inputs.
b/ On a cash basis until 1988, accrual thereafter.
c/ Revenues from transfers received from the Central Government are not included.
p/ Preliminary estimates.

CORPOSANA

3.75   The Water and Sewage Company (CORPOSANA) continues to face financial difficulties despite recent improvements. CORPOSANA generates a very small saving, well below its capital expenditures. A substantial increase in the investment that began in 1990 is likely to enlarge the deficit further.

3.76   Nonetheless, preliminary estimates show some bright spots. Reflecting increases in efficiency in collections, and despite fast declining real tariff rates (tariffs went up only 25 percent in 1989), the value-added generated by the enterprise kept pace with GDP growth in 1989. Interest on the external debt rose sharply, largely explaining the decline in current savings (see Table 3.11), With the ten percent increase in tariffs in the first semester of 1990 and further adjustments in the second semester, current revenues and savings recovered relative to GDP. The 50 percent tariff increase in the last quarter of 1990 offset some of the rise in investment expenditures in 1991.

3.77   Despite the adjustments in 1990, water charges remain low by international standards, which encourages wastage and investments larger than necessary; also sewage charges are too low (5 percent of water charges), which further strains CORPOSANA's finances. Since water charges are linked to properly metered consumption, the tariff level and the effectiveness of collection are important issues. This latter problem is compounded by the public sector (including public entities), which pay no consumption charges.

3.78   CORPOSANA's investment program contemplates works worth US$75-80 million over 40 months beginning in 1990. The plan is to increase the water supply to Asuncion by 70 percent (from 200 to 341 cubic meters). The program includes bringing water from the basic sources to reservoirs, constructing a treatment plant, and constructing a distribution network to consumers. The program will be financed by IDB (US$48 million), a French consortium (US$20 million), and the local counterpart (US$8 million).

Table 3.11: PARAGUAY - CORPOSANA FINANCIAL ACCOUNTS
(% GDP)

	1980/81	1983/84	1986/87	1988	1989	1990 p/
Value added a/	0.19	0.15	0.16	0.16	0.17	0.24
Current Expend.	0.12	0.12	0.12	0.09	0.13	0.16
Wages	0.10	0.10	0.07	0.07	0.08	0.12
Interest b/	0.02	0.02	0.05	0.02	0.05	0.04
Taxes	0.00	0.00	0.00	0.00	0.00	0.00
Other	0.00	0.00	0.00	0.00	0.00	0.00
Own Savings	0.06	0.03	0.04	0.07	0.04	0.08
Real Invest.	0.35	0.30	0.35	0.23	0.25	0.60
Other	0.00	0.00	0.02	0.02	0.00	0.00
Deficit b/ c/	0.29	0.27	0.33	0.18	0.21	0.52

Source: Central Bank, Ministry of Finance, and World Bank estimates.

a/ Includes other own revenues but it is mainly value of sales less cost of intermediate inputs.
b/ On a cash basis until 1988, accrual thereafter.
c/ Revenues from transfers received from the Central Government are not included.
p/ Preliminary estimates.

3.79   Although the envisaged investment program may deserve a high priority, it would be prudent to increase water and especially sewage tariffs in order to control future demand, rationalize future investment outlays, and improve the enterprise's financial performance. The public sector should pay for its consumption. Overstaffing (100 to 120 connections per worker) also is a problem, along with inadequate organization and management practices. An IDB loan includes technical assistance for institutional strengthening, with special emphasis on improvements in the organizational structure and in the use of labor (i.e., retraining of workers and cuts in employment for those who cannot be trained).

FLOMERES

3.80   FLOMERES is Paraguay's riverway transport company. In a landlocked country strongly dependent on river transport, FLOMERES has been associated with national security. The firm is small compared to other PEs and although its finances do not impose a heavy burden on the nation, its economic performance is weak. Regulations require that 50 percent of the country's river transport be carried out by Paraguayan ships. In practice the proportion never reaches 30 percent--a little less than half of it provided by FLOMERES, and the rest by private companies. FLOMERES management complains that it is difficult to compete with Argentina's highly subsidized river vessels, but poor investment choices (for example, buying ocean passenger ships instead of river barges) seem to be far more relevant issues in explaining its financial difficulties.

Table 3.12: PARAGUAY - FLOMERES FINANCIAL ACCOUNTS
(% GDP)

	1980/81	1983/84	1986/87	1988	1989	1990 p/
Value added a/	0.03	0.06	0.12	0.11	0.04	0.07
Current Expend.	0.04	0.04	0.06	0.07	0.13	0.15
Wages	0.04	0.04	0.03	0.03	0.04	0.05
Interest b/	0.00	0.00	0.03	0.04	0.09	0.10
Taxes	0.00	0.00	0.00	0.00	0.00	0.00
Other	0.00	0.00	0.00	0.00	0.00	0.00
Own Savings	-0.01	0.02	0.06	0.04	-0.09	-0.08
Real Invest.	0.02	0.45	0.05	0.06	0.01	0.01
Other	0.00	0.00	0.00	0.00	0.00	0.00
Deficit b/ c/	0.03	0.43	0.00	0.02	0.10	0.09

Source: Central Bank, Ministry of Finance, and World Bank estimates.

a/ Includes other own revenues but it is mainly value of sales less cost of intermediate inputs.
b/ On a cash basis until 1988, accrual thereafter.
c/ Revenues from transfers received from the Central Government are not included.
p/ Preliminary estimates.

3.81   FLOMERES' value-added as a share of GDP plummeted in 1989 because PETROPAR stopped subsidizing the company's bunker oil and gas transport, which was the main nonseasonal transport business in the country. Reflecting lower value-added, savings dropped correspondingly (see Table 3.12). The Government has assumed the service of the company's external debt (about US$90 million). Investment was cut drastically to less than a sixth of 1988, which contributed to hold the deficit, excluding interest payments, to levels similar to those of prior years (as a percentage of GDP).

3.82   This performance was repeated in 1990, when investment was again nil and the enterprise could barely finance wages paid. FLOMERES management agrees that private sector participation in the company may be required, but thinks that the Government must keep some significant participation (a veto power, for example) to protect national security interests. Although authorities in the Ministry of Transport agree with this view, privatization is an option that should be carefully evaluated and implemented if it is the most efficient alternative.

3.83	The Ministry of Transport and Communications in connection with Japan's Technical Assistance Agency (JICA) is completing a global sector study of the Paraguayan transport sector, including waterway uses. This is an important endeavor, since Paraguay's transport costs seem exceedingly high. River transport is expected to be more expensive than ocean transport originating in deep-water ports, but the difference seems too high along Paraguayan rivers. Exporters often prefer to ship from the deep-water port of Paranagua (near Santos in Brazil) paying the corresponding ground transport costs to get there (usually very high), than to ship from Buenos Aires or other River Plate ports using river transport to reach these other ports.

INC

3.84	The cement company (INC) expanded substantially after 1982, even though Itaipu's main construction works were complete. As a result the company is greatly overdimensioned for the Paraguayan market. The expansion was financed abroad, mainly through a French consortium. In addition, the Paraguayan Authorities allege that the real value of the works carried out to be far less than the amounts committed. They may contest this debt in court.

3.85	Production capacity is 1.06 million tons a year while output was just 340,000 tons in 1989. Export prospects to Brazil and Argentina are not good but there may be a potential market in Bolivia. The effects of the unneeded expansion in production capacity are reflected in Table 3.13. Investment was high between 1983 and 1988, the enterprise's value-added has not increased correspondingly. Moreover, the wage bill increased along with capacity not with sales; thus savings, which were positive in the early 1980s, became strongly negative after 1985. Although the deficit was reduced in 1989, the improvement does not reflect productivity increases but the completion of the works that were overdimensioned from the start. The deficit was reduced further in 1990 as a result of further cuts in investments reflecting completion of ongoing works. Investment only included mining equipment and trucks.

3.86	There seems to be much room for improving INC's use of resources. Employment could be almost halved--efforts are underway to reduce staff to 700 from the current 1,100; transport costs may also be reduced; and, as indicated, the authorities are considering litigation in some debt service payment--if settled in favor of Paraguay, costs would be cut significantly. INC also has a G56 billion debt with the Central Bank which it is not servicing. Except for external and internal debt service, it needed no financing from the banking system in 1989-90.

3.87	As for privatization, the Ministry of Industry and Commerce (MIC) tried to implement an ambitious plan. This was delayed (if not abandoned) because of strong political resistance. At a minimum, joint ventures should be explored.

Table 3.13: PARAGUAY - INC FINANCIAL ACCOUNTS
(% GDP)

	1980/81	1983/84	1986/87	1988	1989	1990 p/
Value added a/	0.20	0.25	0.19	0.27	0.31	0.45
Current Expend.	0.06	0.20	0.25	0.39	0.70	0.83
Wages	0.05	0.06	0.08	0.08	0.10	0.12
Interest b/	0.00	0.13	0.17	0.31	0.60	0.71
Taxes	0.00	0.00	0.00	0.00	0.00	0.00
Other	0.00	0.00	0.00	0.00	0.00	0.00
Own Savings	0.14	0.04	-0.06	-0.12	-0.38	-0.38
Real Invest.	0.14	1.83	1.03	1.04	0.19	0.02
Other	0.00	0.02	0.00	0.00	0.00	0.00
Deficit b/ c/	0.00	1.81	1.09	1.17	0.57	0.40

Source: Central Bank, Ministry of Finance and World Bank calculations.

a/ Includes other own revenues but it is mainly value of sales less cost of intermediate inputs.
b/ On a cash basis until 1988, accrual thereafter.
c/ Revenues from transfers received from the Central Government are not included.
p/ Preliminary estimates.

LAP

3.88   The operating results of LAP, the national airline, deteriorated significantly in 1989-90, which was partially offset by larger transfers from the Central Government. In 1988, LAP's current savings were negative, -0.1 percent of GDP, more negative in 1989, and even more so in 1990 (-0.4 percent of GDP) due to collapsing revenues (value-added in Table 3.14). There has been some improvement in management of resources: LAP now shows quarterly balances, has improved the use of its equipment by flying less, and changed the aircraft maintenance program, which used to be provided by Israel Aircraft Industries but now is done with the enterprise's own resources, with the help of foreign experts. Additional management improvements should have taken place in 1990 as a result of recommendations from an external audit by Price Waterhouse. However, results have been disappointing. Sales declined in 1989 and have remained low. The wage bill declined slightly in 1989 (relative to GDP) but rose dramatically in 1990 as did the costs of inputs (basically fuel).

Table 3.14: PARAGUAY - LAP FINANCIAL ACCOUNTS
(% GDP)

	1980/81	1983/84	1986/87	1988	1989	1990p/
Value added a/	0.12	0.30	0.27	0.30	0.14	0.07
Current Expend.	0.17	0.32	0.36	0.42	0.36	0.50
Wages	0.14	0.24	0.30	0.32	0.28	0.40
Interest b/	0.03	0.08	0.04	0.06	0.04	0.07
Taxes	0.00	0.00	0.00	0.00	0.00	0.00
Other	0.00	0.00	0.02	0.03	0.03	0.03
Own Savings	-0.05	-0.02	-0.09	-0.11	-0.22	-0.44
Real Invest.	0.09	0.22	0.28	0.27	0.31	0.12
Other	0.00	0.00	0.00	0.00	0.00	0.00
Deficit b/ c/	0.15	0.24	0.37	0.38	0.53	0.55

Source: Central Bank, Ministry of Finance, and World Bank estimates.

a/ Includes other own revenues but it is mainly value of sales less cost of intermediate inputs.
b/ On a cash basis until 1988, accrual thereafter.
c/ Revenues from transfers received from the Central Government are not included.
p/ Preliminary estimates.

3.89   Large Government transfers have not been enough to finance LAP's capital expenditures; thus, its accounts have significant deficits. Due to lack of resources stemming from poor financial and managerial performance, and given the lack of progress from ongoing adjustments and reforms, LAP should contemplate no significant investment for coming years (this did not happen in 1989 and preliminary estimates show little improvement in 1990).

3.90   LAP's strategy is to strengthen the company's regional position and emphasize the European market, given the opportunities that its deregulation will open after 1992. However, these goals require adequate aircraft and the present fleet is obsolete. LAP has attractive landing rights but capital requirements will be heavy and resources are scarce. The present management would like to find a suitable partner abroad. It dislikes the idea of outright privatization, but welcomes the possibility of joint-venture arrangements with international companies. Discussions in this regard are evolving. Another option would be to sell off the existing assets and pursue an "open skies" policy.

ACEPAR

3.91   ACEPAR, the national steel company, is a mixed enterprise: 74.6 percent is owned by the Ministry of Defense (previously through a now-divested holding company, SIDEPAR) and the rest by two Brazilian enterprises (24.4 percent by FLN, a holding company, and 1 percent by Tenenge, an engineering consulting firm). The company has a technical production capacity of 175,000T cast iron, 240,000T steel, and 120,000T sheet steel. It began full operations in 1988, operating since then with only one of its two blast furnaces and keeping the other idle.

Table 3.15: PARAGUAY - ACEPAR/SIDEPAR FINANCIAL ACCOUNTS
(% GDP)

	1987	1988	1989	1990 p/
Value added a/	0.01	0.06	0.15	0.20
Current Expend.	0.02	0.07	0.24	0.25
Wages	0.02	0.07	0.07	0.06
Interest b/	0.00	0.00	0.17	0.19
Taxes	0.00	0.00	0.00	0.00
Other	0.00	0.00	0.00	0.00
Own Savings	-0.01	0.00	-0.10	-0.05
Real Invest.	0.00	0.00	0.00	0.03
Other d/	0.50	0.75	0.13	0.00
Deficit b/ c/	0.51	0.75	0.23	0.08

Source: Central Bank, Ministry of Finance and World Bank calculations.

a/ Includes other own revenues but it is mainly value of sales less cost of intermediate inputs.
b/ On a cash basis until 1988, accrual thereafter.
c/ Revenues from transfers received from the Central Government are not included.
d/ Reflects real investment before 1989.
p/ Preliminary estimates.

3.92   ACEPAR nearly eliminated its deficit in 1989 (excluding interest payments), down from 0.75 percent of GDP in 1988, as its primary works were completed and sales increased. Most of ACEPAR's external debt was with Brazil, which was rescheduled under very favorable conditions and was fully repaid in 1990--thus the enterprise now owes a far lower debt (about one-fourth of the original contracts), this time with the Government.[45] However, these accounts overstate the viability of the enterprise because potentially serious long-term environmental problems have not been considered. ACEPAR's furnaces are fueled by charcoal made from wood and the costs of a fast dwindling forest have not been included appropriately in the enterprise's financial picture. Estimating these costs is urgent, and if high enough the alternative of closing ACEPAR will need to be examined.

3.93   An encouraging aspect is that 60 percent of the enterprise output is exported, despite working at 50 percent capacity (it uses only one blast furnace). According to its authorities, if the enterprise could get money to begin operating its unused furnace (at an initial investment cost of about US$3 million), it would be able to produce higher quality steel, thus improving its financial outlook. But economic studies on the profitability of firing the second furnace are not available. It is not clear that there will be a market for the additional output and as stated above, important costs have not been estimated.

3.94   ACEPAR management would like to increase capital by issuing new stock without concern about who buys the new shares. The recent divestiture of SIDEPAR should facilitate this process. The Ministry of Industry and Commerce (MIC) believes that Taiwan may be interested in buying a share of the enterprise. Argentine and Uruguayan concerns have also expressed similar interest. If ACEPAR turns out to be socially profitable, export prospects should be studied and joint ventures with the private sector encouraged to increase efficiency and availability of funds.

PETROPAR

3.95   PETROPAR refines and distributes oil and gas. As the principal beneficiary of the subsidized official exchange rates in the 1980s, PETROPAR accumulated surpluses of over US$50 million towards the construction of a new refinery. In 1989, the foreign exchange subsidy was eliminated and taxes were raised[46] but domestic fuel prices were allowed to increase only 15 percent, well below inflation. In addition, the price structure was wrong: gasoil prices were especially low relative to other fuels. Although financial controls were tightened significantly and important cost-savings were made--the premium on the Sahara blend crude that Paraguay imports, was cut from US$1.00-US$1.50 to just 40 cents per barrel, maritime freight costs declined, and fluvial freight cost were cut by eliminating the subsidy granted to FLOMERES of about 85 cents per barrel--PETROPAR's current savings decreased dramatically. However, about 60 percent of the increased deficit reflected higher taxes.

3.96   PETROPAR also increased its investment substantially in 1989. In response to the consolidation of PETROPAR's large deposits in the Central Bank at no interest, PETROPAR tripled its fuel inventories, from 30 to 90 days of consumption. Although 90 days inventory may be too much in social terms, the increase was not the result of a "socially" profitable investment but a "private" decision to reduce deposits because the enterprise could not

---

[45]   Repayment of the debt with Brazil was recently completed, but by the Central Bank, not ACEPAR, at a 70 percent average discount (para 1.53). ACEPAR now owes the Central Bank $10.5 million a year in interest.

[46]   PETROPAR has been the second largest "tax collection agency" in the country after customs.

manage them and earned no interest. This large increase in investment, on top of the sharp fall in current savings, turned PETROPAR's 1988 surplus into a large deficit.

Table 3.16: PARAGUAY - PETROPAR AND APAL FINANCIAL ACCOUNTS
(% GDP)

	1983/84	1986/87	1988	1989	1990p/
Value added a/	0.40	1.56	1.31	0.99	2.11
Value of sales	3.72	4.36	4.24	3.25	5.14
Intermediate inputs	3.32	2.80	2.93	2.26	3.03
Current Expend.	0.32	0.92	0.70	1.07	1.37
Wages	0.13	0.11	0.12	0.09	0.09
Interest b/	0.13	0.03	0.03	0.01	0.01
Taxes	0.05	0.78	0.55	0.96	1.25
Other	0.01	0.01	0.01	0.01	0.01
Own Savings	0.08	0.64	0.61	-0.08	0.74
Real Invest.	0.18	0.53	-0.01	0.57	0.01
Other	0.00	0.02	0.03	0.16	0.00
Deficit b/ c/	0.10	-0.09	-0.58	0.81	-0.73

Source: Central Bank, Ministry of Finance, and World Bank estimates.

a/   Includes other own revenues but it is mainly value of sales less cost of intermediate inputs.
b/   On a cash basis until 1988, accrual thereafter.
c/   Revenues from transfers received from the Central Government are not included.
p/   Preliminary estimates.

3.97   The situation improved significantly in 1990, despite the rise in crude international crude oil prices. PETROPAR raised prices substantially and remedied the structural problems. In January, prices of alcohol for fuel uses, liquified gas and jet fuel were freed--kerosene prices were freed in April; PETROPAR's monopoly on imports of these products was also relinquished. Although gasoline and gasoil prices remain controlled (and PETROPAR keeps its monopoly to import them), they were raised 23 percent in January and 46 percent in April. A further 33 percent increase was granted in September, as needed to deal with the effect of the Persian Gulf crisis on petroleum prices. Fuel subsidies on public transportation were also eliminated at that time. Provided these prices can be kept free of government control and adjusted with inflation, the main problem from a national point of view, is the monopoly power that PETROPAR holds on key products. Price increases helped savings go up despite large increases in taxes. Although a small reduction in petroleum prices took place in April/May 1991, PETROPAR finances have remained strong in 1991. Further savings may take place in the future if a recently reached agreement to buy Argentine crude (replacing the Algeria light) delivers on its promised savings.

3.98   PETROPAR's investment program contemplates increasing fuel storage capacity to 120 days of consumption. As expressed in para. 3.95, this may be excessive from the standpoint of the social productivity of investment. However, the Central Government has limited influence on the enterprise's decisions and so the project may be carried out independently from national priorities. Converting the existing refinery to heavier crudes at a cost of US$80-100 million is another potentially large investment in the program; technical studies have been completed for the project, but the economic evaluation is still pending. Works in this project are not programmed for the near future because of lack of resources, but the recent adjustments in fuel prices might accelerate this schedule. The economic viability of any new refinery should be carefully studied before starting the project. The new refinery will be easier to justify if it can compete with alternative sources for similar products--in other words, after eliminating PETROPAR's remaining import monopolies (gasoline and gasoil).

3.99   PETROPAR has taken over the administration of the alcohol plant (APAL), an inefficient endeavor worth no more than US$4 million but holding US$22 million in liabilities. It carries an external debt of US$9 million and a domestic debt (with the Central Bank) of G13 billion. PETROPAR is proposing to take over APAL permanently while assuming the external but not the domestic debt; however, no final decision has been made yet. The firm plans to improve the efficiency of the plant by making use of byproducts now wasted--a gas that can be used in sodas; vinaza (wine dregs), a potential fertilizer from sugarcane; and the "bagazo," which can be used as animal feed when mixed with vinaza and yeast.

Summary

3.100   Public enterprise finances generally deteriorated in the 1980s, with their total cash primary deficit reaching close to 3 percent of GDP--and much more than that if interest arrears are taken into account and revenue from the foreign exchange subsidy is disregarded (expenditures in foreign currency are valued at an exchange rate that excludes the subsidy). At the core of the PE's financial problems and inefficiencies were lack of managerial accountability and poor management, overdimensioned investments, and overstaffing. Low tariffs contributed to the poor financial performance of public enterprises in the 1980s, but the adjustment that took place in 1989 and especially in 1990, relieved this problem (except for CORPOSANA--water and sewage). Although it is important to develop a mechanism to ensure regular adjustment of tariffs, a danger also exists that public enterprises may take advantage of their monopoly position, raising prices to hide inefficiency beneath an apparent financial strength, but at the cost of reducing the country's competitiveness abroad.

3.101   Positive developments that took place since 1989 included cuts in investment (linked among others to the completion of ongoing works of doubtful economic value--a cement plant and a steel mill), and management improvements. Overdimensioned investments are in most cases a foregone cost; the concern is rather to avoid further repetitions. Completion of some endeavors has permitted cuts in capital expenditures in the cases of INC and ACEPAR, with such expenditures now being kept under tight control in most enterprises (capital outlays also dropped substantially in ANTELCO and FLOMERES). However, because of a strong increase in investment by PETROPAR, ANDE and CORPOSANA, the aggregate investment of PEs increased from 3.6 percent to 6.6 percent of GDP. There is substantial room for better uses of available real and financial resources. Plans to reduce overemployment are in place and implementation has begun in INC and ACEPAR, and are expected to extend to CORPOSANA, LAP, and ANTELCO. Authorities are also planning to improve control and accountability in PE management; also, audit processes are in progress in several enterprises.

3.102   As for management improvements and better use of resources, privatization is a promising option but it has slowed after confronting serious political obstacles (INC), although SIDEPAR was eliminated and some privatization initiatives are taking place in other entities (APAL, LAP, and possibly ACEPAR and FLOMERES). Other actions contemplated along with privatization include normalizing domestic and foreign obligations and modifying some public enterprises' legal status.

3.103   Notwithstanding these improvements, much is yet to be done. The institutional setup under which public enterprises operate is far from satisfactory. This makes control of their efficiency and investment program difficult. Pricing practices must be watched closely--prices must be adjusted with inflation and should not be allowed to decline so as to endanger enterprises' finances, but neither should they be allowed to increase too much and thereby hide inefficiencies and reduce competitiveness abroad and/or welfare among final users of public services. The Government plans to expand its enterprises that provide basic goods and services such as water, electricity and telephones. However, it will be important to not only increase their resources (labor, use of inputs, and investment) but to improve their efficiency. The Government also plans to retreat from money losing activities such as steel, cement, and alcohol, but privatization has not gone very far. Privatization and joint ventures are important options for improving efficiency, and the Government has started a new program to divest some inefficient public enterprises.

D. Tax Reform

3.104   It is often argued that Paraguay's tax system has not kept pace with domestic inflation and growth, which endangers macroeconomic equilibrium, future growth, and prospects for eradicating poverty. Several new taxes were introduced in 1967-69, which <u>temporarily</u> increased revenues (see para 1.10). Tax revenues rose to 10.3 percent of GDP in 1970 (see Table 3.17). However, excluding the transitory jump in revenues in 1968-70, taxes remained in the range of 8-9 percent of GDP in the 1960s and 1970s.

3.105   A fall in revenues took place in the 1980s--taxes dropped to about 7 percent of GDP and remained there until 1988. This was mainly because of a drop in import tax collection. To a large extent these lower revenues from import taxes were the outcome of the system of multiple exchange rates. Import taxes represented almost 3 percent of GDP in 1977/79 and less than 1 percent after 1983 when several exchange rates were used and import taxes were linked to the lower rates.[47]

Table 3.17: PARAGUAY - TAX BURDEN, 1970-89
(in Billions of Guaranies)

	Tax Revenues	GDP	Tax Burden (%)
1970	7.7	74.9	10.3
1971	7.8	83.7	9.3
1972	7.8	96.9	8.1
1973	10.0	125.4	8.0
1974	14.2	168.0	8.5
1975	15.5	190.4	8.1
1976	16.7	214.1	7.8
1977	23.0	263.6	8.7
1978	30.2	322.5	9.4
1979	38.7	430.5	9.0
1980	45.6	560.5	8.1
1981	51.0	708.7	7.2
1982	58.6	737.0	7.9
1983	55.6	818.1	6.8
1984	71.7	1070.4	6.7
1985	96.3	1393.9	6.9
1986	125.0	1833.8	6.8
1987	177.9	2493.6	7.1
1988	233.0	3319.1	7.0
1989	406.6	4608.4	8.8
1990	596.2	6474.4	9.2

Source: Central Bank, Ministry of Finance, and World Bank estimates.

3.106   A significant recovery in tax collection took place in 1989. However, the tax system remains unsatisfactory. The system is based on multiple specific taxes with widespread exemptions instead of a few generalized ad-valorem ones with lower rates; outdated bases for property taxes; and payments made with long delays and no practical penalties. Tax administration also is a big

---

[47] Figures for foreign trade taxes (mainly imports) in this section and in the section on Trade show some difference. For purposes of consistency in the series, only officially labelled import taxes are included here; in the trade section these figures were sometimes adjusted to include revenues that are legally labelled internal taxes but are collected by customs, for example in Tables 3.3 and 3.4.

problem. The cadastre for taxing properties is obsolete and needs updating. Until recently, there were several independent tax units in charge of similar tasks, which created confusion and waste. It is urgent to carry out a profound tax reform, and the Government has concrete ideas on what to do and how to do it. At present, its efforts to increase revenues are two pronged: to improve tax administration under the existing structure, and to have a tax reform approved and implemented by 1992.

3.107   Paraguay's tax/GDP ratio is among the lowest in the hemisphere. Although this implies that the tax burden of the private sector is low, the limited volume of public sector resources also has some drawbacks: public sector wages are low (which encourages corruption); public investment in some of the standard public sector areas such as transport, basic health, and education has been limited; and current social expenditures benefitting the poor are low. And, when the resource constraint was neglected, inflation increased and balance of payments difficulties developed. In the future, Paraguay will receive additional resources from Itaipu (see 2.37 - 2.44), and some capital resources as the privatization process proceeds (see paras. 3.102 - 3.103). Nonetheless, further efforts will be needed to improve the tax system, not just to increase collections but to improve efficiency by reducing evasion and widening the base of more modern new taxes, as explained below.

The Present Tax System

3.108   Taxes under the present system can be classified into four broad categories: taxes on goods and services, income taxes, taxes on capital, and foreign trade taxes. The latter were discussed in the section on trade policies.[48] The first category groups sales tax, several selective consumption taxes (fuels, liquor, cigarettes, livestock, etc.), stamp taxes on different kinds of transactions, and several other small taxes. The overall income elasticity of the aggregate taxes collected in this category has been close to one in the last decade. They represented about 4.2-4.3 percent of GDP in 1984-88 (Table 3.18). However, because of an increase in stamp taxes in 1989, the aggregate grew to 4.6-4.7 percent of GDP in 1989-90.

3.109   The structure of taxes on goods and services has changed significantly. The general sales tax represented 0.6 percent of GDP in 1984, a ratio that increased progressively to 0.8 percent by 1987. Although exemptions are widespread and potential revenues are difficult to calculate, evasion must be widespread, since the tax rate on domestic sales is 4 percent, and on imports 8 percent or 14 percent, with 80 percent of the proceeds coming from the latter two (therefore the sales tax basically amounts to an imports tax). The domestic sales tax is charged only to the final consumer. Even if it affects only half of GDP, it should generate more than 2 percent of GDP in revenues but collections are less than half of that.

3.110   Selective consumption taxes have declined from about 1.6-1.7 percent of GDP to about 1.4 percent of GDP. Stamp taxes, on the other hand, have shown an elasticity higher than one; they were 1.8 percent of GDP in 1984, 1.9 percent in 1988, and 2.4 percent in 1989-90 (see Table 3.18). Though all these taxes are classified under one heading, they include many taxes with no relation to each other; for example, the stamp tax includes 84 different taxes affecting

---

[48] See previous footnote on the categorization of these taxes into "internal" taxes and import duties.

civil and commercial dealings. Many of these taxes are specific and thus decline in importance with inflation. Erosion of potential tax revenues also occurs through widespread exemptions.

3.111   The second broad category of taxes, income taxes, is another collection of uncoordinated small taxes. They apply mainly to enterprises, since the personal income tax applies in few instances and is negligible in effect. The income tax on profits (agriculture is exempted) is slightly progressive, with rates moving from 25 percent to 30 percent with higher corporate incomes. Income taxes represented 1.6 percent to 1.7 percent of GDP in the early 1980s, but this share dropped to 1.2 percent in 1984-86; they were about as much in 1990. Evasion must be pervasive in this category as well. With returns to capital amounting to about half of value added and assuming the tax applies to half of GDP, the enterprise tax alone should represent 6.5 percent of GDP, over 5 times the actual collection.

Table 3.18: PARAGUAY - TAX STRUCTURE
(% GDP)

	1984	1985	1986	1987	1988	1989	1990
Taxes on Goods & Services	4.21	4.26	4.18	4.28	4.30	4.76	4.61
Consumption	2.26	3.35	2.40	2.41	2.30	2.16	2.21
General Sales	0.56	0.66	0.74	0.82	0.81	0.79	0.78
Selective Sales	1.70	1.69	1.66	1.59	1.49	1.37	1.43
Stamp Taxes	1.79	1.77	1.65	1.76	1.90	2.44	2.34
Other	0.15	0.14	0.13	0.11	0.10	0.15	0.06
Income Taxes	1.11	1.27	1.25	1.53	1.42	1.40	1.26
Capital Taxes	0.38	0.40	0.38	0.33	0.28	0.25	0.27
Land/Property	0.35	0.37	0.35	0.30	0.25	0.22	0.27
Other	0.03	0.03	0.03	0.03	0.03	0.02	0.00
Other Taxes a/	1.01	0.97	1.01	0.99	1.02	2.42	3.08
Total	6.71	6.91	6.82	7.13	7.02	8.82	9.21

Source: Ministry of Finance.

a/ Mostly import and export taxes.

3.112   Capital taxes generate little revenue--about 3 percent of total taxes and, recently, less than 0.3 percent of GDP. Although tax rates are about 1 percent of property value, assessments are extremely low. On the average, urban property tax values represent less than 35 percent of market value, and in rural areas taxable values are just 5 percent of market value. Moreover, the cadastre is incomplete, with many rural properties never assessed. The capital tax category also includes an inheritance tax that is so easily evaded its proceeds are insignificant.

Other capital taxes included are special taxes on the value of stocks, shares, and silent partnerships, as well as taxes on automobiles, though again revenues are negligible.

3.113   Sanctions for failing to pay taxes due on time vary by tax and often are nonexistent. In fact, the system not only does not penalize infractions, but in practice encourages the taxpayer to avoid payment. In most instances, penalty interest rates are lower than commercial interest rates. Thus, for the taxpayer, it is more profitable to delay payment until the infraction is discovered (if it ever is) than to pay on time.

Ongoing Administrative Reforms

3.114   Recently, the Finance Ministry in general, and tax administration in particular, were substantially reorganized. For administrative purposes, the tax system used to emphasize independent and parallel organizations for each main group of taxes. There were four tax divisions: sales and consumption, income, property and import. Each one had its own independent department of data processing (including different and incompatible computers), tax collection, accounting, auditing, legal advisors, penalties for lack of payments or errors in calculating taxes, etc. The main goal for these divisions was to fulfill targets set in the Budget, which is why underestimating tax revenues in the Budget has been a tradition.

3.115   A new Under Secretary of Taxes was created in the Ministry of Finance, which unified the four tax divisions mentioned. The implementation of an organizational scheme based not on types of taxes but on tax functions is expected soon. Efficiency will be enhanced by eliminating duplication. The new divisions would be, for example, Tax Collection, Fiscalization, Support Services, etc., with Customs also a division under this new Under Secretary.

3.116   Authorities have started to implement the new Single Tax Registry (Registro Unico de Contribuyentes, RUT), approved in December 1988. Taxpayers (persons and legal entities) are being registered under this new registry that will provide information on all tax-related transactions corresponding to each taxpayer, thus facilitating control. The computer system to manage the RUT is already in place. Authorities expect that in two years they will have everyone registered, but admit that the deadline for the important taxpayers already has passed. The delay was related to the elimination of the General Directorate of Taxes (which had functions and purposes very similar to the new Under Secretary of Taxes), created in the previous administration together with the RUT. The new Under Secretary is expected to expedite registration.

The Present Tax Reform

3.117   Authorities have recently proposed a plan to reduce the number of taxes to 11 while making the system simpler and easier to manage. Of those 11, about 9 taxes are expected to evolve from existing levies, though substantially modified. A large effort is contemplated to eliminate taxes that are cumbersome and yield little revenue; also, specific taxes will be transformed into ad-valorem ones, and tax assessments will be updated. In addition, a new simple value-added tax is planned.

3.118   In the proposed reform, the system of indirect taxes will be based on: (i) a value-added tax (VAT) that will replace existing sales taxes; and (ii) a few selective ad-valorem taxes on consumption (fuels, liquors, cigarettes, luxuries, etc.) that will replace specific taxes now in place as well as the important stamp tax. A flat rate (to be negotiated in Congress) and a wide base are being considered for the VAT, but many exemptions are still contemplated. The VAT will operate as a normal sales tax, but the taxpayer will be able to deduct from his taxes the VAT that he pays on his inputs. Agriculture is exempt and some public enterprises may end up exempted too. Agriculture's exemption is only for first-round transactions, which is considered efficient because those transactions would be too difficult to monitor. Further transactions of agricultural inputs involving the industrial sector would be taxed. Because of monitoring problems, small enterprises (those with income below a certain level) will be exempt, but a Unified Tax will be charged on them, replacing the VAT as well as direct taxes they might owe. This tax would be based on easy-to-identify indices, and allow beneficiaries to deduct a fraction of the VAT that they in turn had paid. In this way, authorities expect that large enterprises will have trouble evading the VAT through bogus transactions with small enterprises.

3.119   Initially, a VAT with a hypothetical 12 percent rate would be expected to yield 2 percent of GDP but would replace taxes that contribute 1.6 percent; thus, the net impact would be an increase of 0.4 percent of GDP (see Table 3.19). It would apply to about 30 percent of GDP and to most imports; this estimate assumes that evasion would remain significant, at least in the early stages of implementation. It is fundamental to negotiate lower rates in exchange for a wider base.

3.120   For selective consumption taxes, ad-valorem rates will replace specific rates; the ad-valorem rates are expected to have a ceiling established by law and defined for each tax, but authorities will be able to decide the actual rate up to that ceiling. Through this process, authorities intend to increase tax elasticity while reducing the discretion now in the hands of tax collectors, an apparent source of corruption under the present system.

3.121   For income tax on profits, the proposal is to tax them at a flat rate also to be negotiated. Other reforms include proper assessment of property values[49] based on a comprehensive cadastral study that may take a number of years to complete. The inheritance tax would be abolished, and the tax on vehicles is intended to tax use, not property to avoid the problem posed by the large number of vehicles where legal ownership is difficult to ascertain.

3.122   The Authorities are discussing the Tax Code with penalties and a system of enforcing penalties as separate items. They intend to prepare a law dealing with tax penalties in general instead of the multiple laws and regulations that currently exist. A preliminary draft of this law is now ready, but further improvements are contemplated. It will be critical to include a true penalty interest rate to encourage timely tax payments.

---

[49]   Automatic general revaluation of urban and rural and values can be done by decree and needs no special law. The text refers to proper valuation of land on individual properties.

Table 3.19: PARAGUAY - VAT, ESTIMATED NET REVENUES
(Initial years, % GDP)a/

GDP	100.00
Fraction Taxed	27.28
Fraction Exempt b/	72.72
Imports	6.95
Fraction Taxed	6.17
Fraction Exempt c/	0.78
Potential Revenue	4.01
Expected Initial Evasion	2.00
VAT Collected	2.01
Taxes Forgone	1.65
Net VAT Effect	0.36

Source: Central Bank, Ministry of Finance, and World Bank estimates.

a/   Estimates based on 1988 GDP data.
b/   Includes exports, agriculture, electricity, water, small enterprises, Government services, housing, health, education, and changes in stocks (which cannot be taxed in a system that uses sales as the base).
c/   Oil imports, included in the import total used, are expected to be exempt.

3.123   The Authorities presented a complete draft of the tax reform in August 1990. After discussions with the private sector and the community in general, a project was sent to Congress where it awaits approval. The private sector is expected to be more formally involved in the preparation of the final project. The tax reform then goes to Congress where a strong debate is expected. A two-part strategy is expected to be used in Congress to have the reform passed into law. First would be to approve less controversial modifications to existing taxes; second would be the approval of the new taxes contemplated. Authorities expect that this may take some time; but for the reform to succeed, it is extremely important to have it backed by the private sector. The delay could be useful if it helps increase support by labor and entrepreneurs.

Reflections on a Tax Reform

3.124   The private sector claims that taxes are widely evaded in Paraguay because of what it considers unreasonably high tax rates: the 30 percent tax on profits, some high import tariffs, and the stamp tax (an inefficient scheme that levies contracts rather than output, income, or wealth). To avoid these taxes, the private sector has developed a complex "parallel economy" with surprisingly favorable results relative to those obtained in the "formal economies" in neighboring countries. Whether the private sector is correct or not, experience suggests that Paraguay needs an efficient tax system that is perceived as fair. In setting tax rates, the Government must consider not only their potential revenues or redistributive effects but its ability to enforce the law and collect the taxes due. It is not advisable to set high tax rates that will be evaded and "punish" the few that decide to abide by the law. For a country with a small public sector and Government Authorities that resist enlarging it, the door is wide open for promising and sweeping changes to cut evasion by enacting a few simple taxes with broad bases (no exceptions if possible) and low rates.

3.125   Of course, it would take time and good tax administration to reach potential yields, but with low tax rates, wide bases, few exemptions, and good administration the Government should be able to obtain a reasonable volume of resources. Tax administration and proper backing from the community (both Authorities and the private sector) are indispensable. Some aspects may take time before implementation is possible, but efficient temporary alternatives can be formulated. For example, a cadastre would be an urgent task to carry out. While this is being done, some kind of voluntary valuation of capital can replace the official one--penalties can be devised for gross violations, e.g., valuation can be public, with the Government having the right to purchase property from a taxpayer at the price he values it for tax purposes.

CHAPTER IV: ECONOMIC PROSPECTS

Summary Background

4.1   Paraguay's growth and financial performance were moderately successful in the sixties. Growth rates of about 4-5 percent per year exceeded population growth (slightly less than 3 percent per annum); inflation, at less than 5 percent per year, was similar to industrial countries' rate of 3 percent. Growth accelerated substantially in the first half of the 1970s with the construction of Itaipu (beginning in 1974). Between 1973 and 1981 the annual growth rate exceeded 9 percent per year. Domestic inflation increased reaching annual rates of 10-15 percent--inflation in industrial countries was similar until 1978 but much lower in 1979-81. The large domestic investment surge that occurred during the period of Itaipu construction stimulated domestic demand, production, and employment, and the substantial inflows that accompanied Itaipu were sufficient to keep the nominal exchange rate constant and allow for a substantial increase in foreign exchange reserves. External debt was nearly five times larger in 1981 than in 1973, but due to the high real growth rate, the reduction in the real exchange rate (linked to Itaipú related foreign exchange inflows) and the high international inflation, the ratio of external debt to GDP actually declined slightly over that period.

4.2   Economic policy in the latter half of the 1970s did not prepare the country for post-Itaipu conditions. To a large extent, the country spent the transitory additional resources as if they were permanent: consumption increased and investment projects, particularly construction, assumed unsustainable expenditure patterns. When basic works at Itaipu were completed (1981) and inflows of foreign exchange dropped, expenditures and growth declined. The public sector tried to maintain domestic demand by starting oversized and poorly timed investments, but the result was only to add internal and external financial disequilibrium to the reduction in resources available to the country. Distortions multiplied and productivity suffered. Per capita GDP was 10 percent lower in 1987 than in 1981; the ratio of external debt (medium- and long-term, including arrears) to GDP rose from a low figure of about 20 percent of GDP in 1981 to about 70 percent in 1987-89; the Central Bank lost US$550 million of foreign reserves between 1981 and early 1989; quasi-money in domestic currency dropped from 7.8 percent of GDP in 1981 and almost 10 percent in 1983 to just 4.5 percent in 1990; and inflation doubled to 30 percent in 1986-87.

4.3   At the core of the deteriorating economic performance was a large but hidden public sector deficit and a system of multiple exchange rates. Officially the deficit was less than 3 percent of GDP in 1987, not much larger than in 1981, but the true deficit (including interest arrears and the public sector foreign exchange subsidy) was about 8 percent of GDP. There were seven different exchange rates in 1987, the lowest just 15 percent of the highest (a free rate).

4.4   Radical changes in economic policy took place after February 1989, when new Authorities took charge in the country. The exchange rate was unified. In the public sector, the foreign exchange subsidy was eliminated; a strong program was implemented to rationalize both current and capital expenditures in the Central Government as well as in public enterprises; and tax collection practices were improved. In the financial sector, interest rates were freed, except for rediscounts which still provide a significant subsidy. New savings instruments also were authorized. To lower external debt, a pioneering agreement was reached with Brazil, whereby using Brazilian foreign paper purchased at a discount in secondary markets but valued at par. Using this mechanism the Paraguay's debt to Brazil was prepaid in September 1990.

4.5   The results of these initiatives have been excellent. Between 1988 and 1989, the public sector deficit was cut from 7.6 percent to 3.9 percent of GDP with small surpluses estimated for 1990 and for 1991. Public savings increased more than three percentage points of GDP (if the foreign exchange used to pay for imports and interest is valued at market prices). The balance of payments showed a surplus of over US$200 million between February and December of 1989 and of US$550 million in 1990-91 taken together. Despite tight fiscal policy, real GDP growth averaged close to 4.5 percent per year in 1989-90 well above growth rates of the 1980s. But, notwithstanding much improved public finances, the money supply growth reached record heights in 1989-90, fueling a rise in inflation. Monetary expansion was 15 percent in the 12 months ending in February 1989, and almost 90 percent in the two-year period December 1988-December 1990. The principal source of monetary expansion was the inflows of foreign exchange, which the Central Bank purchased at a roughly fixed exchange rate and added to its international reserves. As the nominal exchange rate has remained almost constant since mid-1989, the real exchange rate has dropped (appreciated) along with domestic inflation.

4.6   The large inflows of foreign exchange in 1990 were related to substantial payments received from Itaipu, overvalued currencies in Brazil and Argentina, and speculative capital inflows. In 1990, Itaipu settled arrears accumulated with Paraguay up to 1989; only the payments corresponding to 1990 were overdue at end-1990. Stabilization programs imposed in Brazil and Argentina in 1990 resulted in large appreciation of these countries' currencies leaving the Paraguayan Guarani highly undervalued with respect to its two large neighbors but highly overvalued with respect to other countries. Speculative inflows developed to take advantage of high dollar-equivalent interest rate in Paraguay, despite the very negative real interest rate when compared to domestic inflation.

4.7   In 1991 the economy largely has overcome these short-run difficulties. The Government has followed a stringent monetary program that has drastically reduced the expansion in the money supply and has cut inflation to annual rates below 15 percent in the first three quarters of the year. The IMF agrees with the program applied and talks towards reaching a stand-by agreement have advanced substantially. However, a final accord has not been signed because of Paraguay's difficulties with the commercial banks.

4.8   Structural reforms also are needed to move the country onto a sustainable growth path. In the financial sector it is necessary to revise complicated regulations and inefficient supervision of institutions, while setting favorable conditions for the recovery of quasi-monetary savings. Progress has been made with the liberalization of interest rates in 1990, a more recent reduction in reserve requirements in 1991, and the implementation of open market operations. In the trade area, it is necessary to simplify customs administration. Tariffs should be adjusted to reach a low uniform rate, in order to bring the customs code into alignment with reality, reduce the potential protection that could develop if the existing custom code were strictly applied, and bring the country into greater harmony with the codes of its MERCOSUR partners. In the public sector, despite the major achievements in 1989, the success is not deeply rooted and old problems may revive. To avoid this, it will be necessary to increase the efficiency of public enterprises. The tax system needs revision to reduce tax rates, widen the bases, and enforce payments, all of which should improve efficiency. To avoid further arrears accumulation and/or unnecessarily heavy adjustment costs, efforts are needed, coupled with cooperation of the international financial system, to resolve Paraguay's debt problems and to resume payments on the external debt.

4.9   Prospects for high growth rates and financial stability are excellent if the short-term and structural issues are dealt with properly. By international standards, many of the adjustment's most painful aspects already have been successfully completed (freeing the exchange rate, liberalizing interest rates, and strengthening public finances). On the other hand, prospects may not be good if the remaining issues are not tackled properly. Three scenarios have been prepared to illustrate the importance of completing the adjustment started and securing international cooperation. A low case shows the consequences of retreating from the adjustment process started. A high case illustrates the effects of rapidly implementing all the changes needed and obtaining the needed funds from abroad. And an intermediate case shows the effects of a slower progress in putting in place the reforms needed while avoiding significant retrogressions. A summary of the main results in the three cases is presented in Table 4.1. Table 4.2 describes the sources and uses of funds needed to finance the balance of payments under the three alternatives.

Low Case: Retreating from the Changes Made

4.10   The consequences of retreating from the reforms already made are serious, since the impressive achievements in the fiscal and external areas would be reversed. This scenario assumes that fiscal policy becomes expansive in 1992 as a result of political pressures stemming from the presidential election in early 1993. Consequently, credit to the Government and monetary growth increase and inflation accelerates. Authorities then reimpose interest rate controls. The tax reform is not approved (or is approved in a form that fails to induce private sector cooperation) and public enterprises' productivity and financial performance deteriorate. As a consequence of the deterioration, the recent speculative capital inflows reverse themselves.

4.11   As noted, the initial effect of an expansive fiscal policy would be higher monetary growth and inflation. In 1992, under this scenario, monetary growth is projected to exceed 37 percent, more than double the 1991 rate. Inflation increases correspondingly, reaching 30 percent. As inflation accelerates, the national currency appreciates and balance of payments problems reappear. At some point in the second semester the real exchange rate need to depreciate would become urgent. In this scenario, Authorities nonetheless are assumed to try to slow down the devaluation rate and to reimpose a system of multiple exchange rates, along the lines of that observed until 1989. Its effect is to deter exports and only gradually increase the implicit exchange rate cost for imports, to about 30 percent in real terms in 1995. This further increases the financial disequilibrium.

4.12   Inflation is expected to accelerate to 30 percent in 1992, double the rate expected in 1991, and would increase further in following years--which makes unorthodox approaches to address the problem politically attractive. The unified exchange rate and freely determined prices are early casualties of such a policy shift.

4.13   Despite the effect of multiple exchange rates, the current account deficit increases. Because foreign creditors are not willing to channel funds to the country, Paraguay cannot finance its deficit with voluntary disbursements, thus it is assumed to accumulate more external arrears as

Table 4.1: PARAGUAY - ALTERNATIVE COURSES OF ACTION

	Historical Background			High Case				Medium Case				Low Case		
	1989	1990	1991	1992	1993	1995	2000	1992	1993	1995	2000	1992	1993	1995
A. International Economic Environment (same for both scenarios)														
Interest Rate (LIBOR)	9.3%	8.5%	8.5%	8.0%	7.6%	7.0%	6.4%	8.0%	7.6%	7.0%	6.4%	8.0%	7.6%	7.0%
International Inflation (<>MUV)	0.6%	6.5%	9.5%	0.8%	0.0%	3.7%	3.7%	0.8%	0.0%	3.7%	3.7%	0.8%	0.0%	3.7%
Terms of Trade a/	113.6	97.9	92.5	94.8	98.1	95.2	95.2	94.8	98.1	98.4	95.3	94.9	98.3	98.7
B. Internal Results														
B1. Growth and Inflation (%)														
GDP Growth	5.8	3.1	3.0	6.3	6.3	6.4	6.8	4.5	5.2	4.7	4.8	4.9	3.9	2.9
Private Investment/GDP	15.4	16.9	17.8	18.0	18.1	19.8	19.8	17.8	17.7	17.8	17.9	17.1	16.5	16.0
Inflation	34.4	36.3	15.0	10.0	4.8	3.7	3.7	13.0	9.0	9.0	9.0	30.0	50.0	70.0
Private Consumption	-5.6	2.0	2.1	5.6	6.0	6.0	5.9	4.4	5.5	4.8	4.4	6.7	6.3	3.6
B2. Private Savings and Credit (% GDP)														
Additional Financial Savings	3.9	0.1	1.3	2.2	2.4	2.2	2.6	1.2	1.4	1.2	1.2	-0.4	-0.3	0.0
Non-financial Savings	17.9	15.2	14.0	13.2	14.1	14.6	15.6	13.7	14.1	14.0	13.9	15.9	14.2	12.5
Inflation Tax	2.6	2.9	1.1	0.7	0.3	0.3	0.3	0.9	0.6	0.7	0.7	2.2	3.6	5.2
Credit to Private Sector (Stock)	11.1	10.6	8.0	11.2	12.9	15.7	21.8	10.6	12.0	14.4	18.8	9.2	9.3	8.4
Domestic Quasimoney (Stock)	4.9	4.5	5.5	6.2	6.9	8.1	11.1	5.7	5.9	6.3	7.3	5.3	5.1	4.9
B3. Public Sector Accounts (% GDP)														
Current Savings	3.0	5.6	5.4	6.5	6.1	6.4	6.5	6.3	5.9	6.1	6.0	2.7	1.8	0.0
Investment	6.8	5.0	5.9	5.2	5.3	5.5	6.0	5.1	5.1	5.2	5.3	4.7	4.5	4.3
Deficit (- = surplus)	3.9	-0.7	-0.5	-1.2	-0.8	-0.8	-0.4	-1.2	-0.8	-0.8	-0.7	2.0	2.8	4.4
B4. External Sector ($mill)														
Exports of G&NFS	1534.6	1805	1872	2090	2347	2887	4806	2038	2244	2653	4023	1952	2042	2146
Resource Balance	241	66	-98	-94	-73	-110	-200	-97	-86	-140	-249	-142	-170	-294
Current Account Balance	218	102	-47	-38	-33	-50	-103	-42	-48	-85	-172	-94	-122	-247
Reserve Accumulation	137	246	300	53	69	95	120	41	57	76	84	20	20	20
Real Effective Exchange Rate	152.8	134.8	142.8	143.8	143.8	143.8	143.8	143.0	143.2	143.4	143.7	135.7	127.5	110.3
Multiple Exchange Rates Effect. b/	0.0	0.0	0.0	0.0	0.0	0.0	0.0	0.0	0.0	0.0	0.0	8.0	12.0	30.0
B5. External Debt and Balance of Payments Financing														
External Debt ($min)	2385	2124	2291	2349	2415	2585	3125	2341	2409	2608	3294	2462	2658	3235
Gross Fund Requirements ($mln)	623.4	590.2	370.7	206.6	210.7	236.4	537.2	198.7	212.7	252.9	577.5	389.8	411.9	528.4
External Debt/Exports (%)	174	142	153	113	103	90	65	115	108	98	82	165	176	216
Debt Service/Exports (%)	41.2	43.3	23.6	14.0	13.1	10.0	11.4	14.4	13.6	10.9	14.0	24.5	23.6	25.1
Interest/Exports (%)	5.0	7.0	7.4	5.1	5.4	4.3	3.0	5.3	5.7	4.7	3.8	7.3	7.5	8.9

a/ International prices are the same in both scenarios; terms of trade differences are due to different compositions of exports and imports.
b/ Exchange rate equivalency of non-tariff restrictions and multiple exchange rates.

it increases import restrictions. In this case, balance of payments troubles arise to some extent because of higher current account deficits but mostly because of difficulties in the capital account. Voluntary gross disbursements from the international financial community decline and private capital outflows become significant, thus, the needed funds are assumed to be provided partly by the accumulation of arrears, and partly by "unidentified" sources.

4.14 Achievements in the public sector are also reversed. The improvements in tax administration evaporate as inflation accelerates and the Tanzi effect overcomes the improvements in tax administration in place. Control of expenditures becomes politically unsustainable, as reductions in GDP growth and employment force the public sector to become employer of last resort. And although not quantified, the increases in efficiency and the better financial picture in public enterprises will likely disappear, as inflation erodes the real value of tariffs charged and political pressures make it more difficult to adjust them enough. The result is that the public sector deficit again increases, to close to 5 percent of GDP in 1995.

Table 4.2: PARAGUAY - SOURCES AND USES OF EXTERNAL FUNDS
(millions of US$)

		High Case				Medium Case				Low Case		
	1990	1992	1993	1995	2000	1992	1993	1995	2000	1992	1993	1995
Total Sources of Funds	345.3	206.6	210.7	236.4	537.2	198.7	212.7	252.9	577.5	389.8	411.9	528.4
Gross Voluntary Disbursements	80.0	715.4	167.8	186.4	279.5	715.4	167.8	186.4	279.5	141.2	127.0	147.8
Direct Investment	73.1	33.3	36.6	46.7	90.3	33.3	36.6	46.7	90.3	33.3	36.6	46.7
Gap	0.0	27.4	6.3	3.2	167.5	19.5	8.4	19.7	207.8	35.3	58.3	83.9
New Arrears	158.2	-569.5	0.0	0.0	0.0	-569.5	0.0	0.0	0.0	180.0	190.0	250.0
Total Uses of Funds	345.3	206.6	210.7	236.4	537.2	198.7	212.7	252.9	577.5	389.8	411.9	528.4
Current Account Deficit	-102.3	38.0	33.3	50.2	103.2	42.3	47.6	85.2	172.3	93.8	122.3	247.3
Amortizations	656.1	185.7	179.2	165.8	403.8	185.7	179.2	165.8	411.1	185.7	179.2	165.8
Private Capital Outflows	-488.4	-70.6	-70.6	-74.5	-89.3	-70.6	-70.6	-74.5	-89.3	90.3	90.3	95.4
Foreign Reserves	245.9	53.4	68.8	94.9	119.6	41.3	56.5	76.4	83.5	20.0	20.0	20.0
Others Not Explained	34.0	0.0	0.0	0.0	0.0	0.0	0.0	0.0	0.0	0.0	0.0	0.0

4.15 Despite a current account deficit that does not explode, this scenario may not offer a sustainable policy option. Private savings decline. (They may even slow more than projected-- in which case inflation would accelerate more than indicated.) Direct investment will not be as large as assumed; GDP growth rate, already in the low side, is likely to be exaggerated; and not even the lower voluntary disbursements assumed may materialize. Social unrest would become widespread. The old problems that developed after 1982 will reappear. For foreign creditors, it would be dangerous to offer the country any significant support. For these reasons, although the tables provided show figures through 1995, it is doubtful that policies under this scenario can be sustained that long.

High Case: Fast Implementing Missing Changes

4.16   The main assumptions used to build this scenario are the following:

(a)   The accumulation of foreign reserves is assumed to be prudent after 1991 (about 3 percent of imports annually), thus allowing for a modest monetary expansion.

(b)   Lending and deposit interest rates remain free (including rediscount credit, which is linked to a free rate). Marginal reserve requirements continue to be cut and/or some interest is paid on deposits kept at the Central Bank (either directly, if legally possible, or indirectly, by creating a new instrument that can only be maintained by those who keep deposits in the CB). These changes should help increase deposit interest rates and reduce lending rates by cutting the spread between lending and deposit rates.

(c)   Tax reform is approved. This should cut tax rates and widen the tax base by eliminating most tax exemptions: For example, a value-added tax with a low rate, no exceptions (although imports are taxed but exports are not), is efficiently enforced; and some simple consumption taxes on luxury goods (such as liquor, cigarettes, cars, motorcycles, sodas, beer) and fuels (to supplement road user fees) replace the myriad of indirect taxes now prevailing. Although some increase in revenues is likely, the key goal would be to increase efficiency.

(d)   A simple 10 percent duty is imposed on all imports, replacing the multiple rates now in place. For simplicity it is assumed that no exemptions are allowed. Exports are exempted from taxes.

(e)   The ongoing studies on ways to improve finances in public enterprises are successfully implemented.

(f)   The international financial community assists Paraguay to resolve the arrears problem and channels the new resources needed to finance its balance of payments.

4.17   The results of these changes are very favorable. Private investment is expected to rise and productivity to increase; thus, GDP growth rates accelerate and remain high. Private investment increases from 18 percent to 20 percent of GDP, and GDP growth rates, after a temporary drop in 1990/91, increase to over 6 percent per year.

4.18   The increase in private investment is more than matched by a rise in private savings, which go up from 15 percent of GDP in 1990-91 to close to 18 percent in the second half of the 1990s. A critical reversal takes place in financial savings, which were consistently negative throughout the 1980s because of negative real interest rates.[50] The interest rate policy plus the reduction in the spread between lending and deposit rates are expected to have a strong effect on the flow of quasi-monetary savings in domestic currency. This quasi-money stock will increase from 4.9 percent of GDP in 1989 to 11 percent by the end of the 1990s, thus exceeding the level they had in 1983 (the flow in real terms as well as the evolution of the stock is given in Table 4.1). Reflecting a higher productivity for capital, nondistributed profits should increase along with nonfinancial savings. These latter rise to more than 15 percent of GDP in the second half of the 1990s.

4.19   Public sector finances remain strong despite conservative assumptions. The small surplus estimated for 1990 continues afterwards. The main sources of strength are a reduction in external interest payments (from 1990 onwards) stemming from the reduction of debt with Brazil, the solution of the arrears problem and the effect of the tax reform that should increase taxes from 9.2 to 10.5 percent of GDP. Public investment is expected to rise slightly as a percent of GDP.[51]

4.20   Inflation declines as a result of limited monetary growth. With public finances in line, strong financial savings and foreign reserve accumulation fully financed, the rate of expansion in the money supply is held to about 10 percent after 1992, enough to keep the inflation rate similar to the international rate (4 percent). Because of the strong public finances and the recovery in quasi-monetary savings, credit to the private sector increases from 9 percent of GDP in 1990-91 to 14 percent in 1993, and to 20 percent in the late 1990s.

4.21   In the balance of payments, the current account deficit drops to about 0.6 percent of GDP after 1993 following a small real devaluation that is expected to take place in 1992, after the real exchange rate increases (the value of the local currency depreciates) in Brazil and Argentina and Paraguayan Authorities succeed in limiting speculative capital inflows. The overall balance of payments shows a modest surplus in 1992 that increases in following years, averaging about US$70-75 million a year between 1992-95.

4.22   After a large reduction in debt in 1990, reflecting prepayment of the debt with Brazil, the external debt in current dollars increases moderately, but falls relative to GDP or exports. Debt was twice the annual value of exports in 1988, but will fall to less than 12 months of exports in the second half of the 1990s. Debt service (and therefore the debt service ratio) drops along with the stock of debt.

4.23   The projections assume significant gross requirements of funds, but only about half those of the previous scenario. Because of a smaller capital outflow than in the Low Case, gross disbursements are estimated at about US$220 million per annum in the first half of the 1990s.

---

[50]   Large increases in M1 (and thus in financial savings) took place in 1989, but this appears to have been only a temporary condition reflecting a money market disequilibrium that was resolved in 1990. Reflecting this adjustment, financial savings were very low in 1990.

[51]   The projection of Government revenues is conservative; it does not include Itaipu royalties and electricity sales to Brazil. If these sizable revenues (3 percent of GDP) materialize, they will need to be managed carefully--Paraguay must avoid another case of the Dutch Disease that occurred in the 1970s.

This modest amount is lower than the country received in 1988/89 but higher than preliminary estimates of disbursements in 1990. Of course, if Brazil cannot fulfill its commitments and therefore Itaipu cannot pay Paraguay for the electricity it sells and the royalties it is entitled to, gross requirements of funds would have to increase correspondingly. The projections assume that arrears will be fully refinanced in 1992, thus opening way for the normalization of Paraguay's relation with its creditors.

Intermediate Case: Slower Progress but no Retrogression

4.24   In this scenario, the pace of reform is slower than under the High Case, reflecting the likelihood of the democratic Congress' lengthy consideration of the measures proposed by the Authorities. On the other, the achievements already made (and assumed lost in the Low Case) are maintained. Assuming the Authorities are able to sustain the several key measures already implemented (in contrast to the Low Case), results are still favorable (although less so than in the High Case). However, because important policy changes are assumed to occur only over time, this scenario may not be quite stable as the High Case. For example, the Authorities might be tempted to follow expansionary policies, along the lines of the Low Case, in which case the latter would better approximate results. Alternatively, they may succeed in implementing the missing reforms rapidly, in which case the High Scenario would illustrate the most likely outcome.

4.25   The Intermediate and High Cases differ mainly in their assumptions about the public sector and in the financial area. In the public sector, no drastic changes take place in the tax area. Tax collection in the Intermediate Case is nonetheless only marginally below that of the High Case because the tax reform is not aimed primarily at increasing revenue but at improving efficiency; the important changes take place in overall productivity, as measured by the ICOR. Current expenditures grows somewhat faster than in the High Case. But because public investment is expected to increase at a slower pace, the overall public sector deficit is similar under the two alternatives.

4.26   In the financial area, it is assumed that Authorities are unable to implement further reductions in reserve requirements and therefore the large differential between deposit and lending interest rates remains. Lower productivity and higher lending rates are translated into lower private sector investment in the Intermediate Case; lower deposit interest rates generate lesser financial savings. Commercial policy lowers tariffs less than under the High Case, thus imports (relative to GDP) are a little lower in the Intermediate Case (which also leads to a slightly lower real exchange rate).

4.27   Nonetheless, results of the Intermediate Case are satisfactory. Tight public finances and a prudent foreign reserve accumulation policy generate a tight monetary policy. Inflation thus comes down, to single digit features, but it exceeds that expected in the High Case. A more cautious response of private investment due to a less plentiful availability of credit (in turn due to the high differential between lending and deposit interest rates) and lower productivity, propels up the economy at a lower speed than in the High Case: growth hovers near 5 percent while it exceeds 6 percent in this latter case.

4.28   In the balance of payments exports grow at a lower pace than in the High Case. Imports also grow less because the reforms simplifying the Customs system do not materialize. These effects more or less balance each other in early years, thus the resource balance is initially similar in the two cases. However, as time goes by, the export effect prevails, which causes a deterioration in the resource balance in the Intermediate Case, relative to the High Case. The current account deficit increases a little faster due to the cumulative effect of a slightly higher debt needed to finance the larger deficit in the resource balance. This makes a gradual depreciation of the local currency necessary. Because of this adjustment, the current account deficit never exceeds 1.5 percent of GDP in the Intermediate Scenario during the projection period.

4.29   Resource requirements are of course higher in the Intermediate than in the High Case, but the differences are moderate. In fact, because of a less aggressive reserve accumulation policy, requirements are lower in 1992 for the Intermediate Case, but this is an exception. On average, requirements are about US$10 million per year higher in Intermediate Scenario in 1993-95, reaching about US$240 million. As in the High Case, this scenario assumes that arrears are fully renegotiated (mostly rescheduled) in 1992. Following this agreement, the projections assume that external financing requirements are voluntarily provided, mostly by multilaterals and bilaterals. Reflecting high amortization payments, requirements become sizable at the end of the decade, but again, this should pose little trouble if the prior disbursements have been properly invested. In fact, as in the High Case, external debt grows moderately, always declining relative to exports or GDP; the debt service ratio follows a similar trend.

Conclusions

4.30   Because of the recent accumulation of foreign reserves, the need to encourage exports and avoid further reductions in the real exchange rate, the availability of money from Itaipu, and the possibility of a favorable outcome of a Paris Club Meeting, Paraguay's most urgent concerns are in strengthening its tax, financial and trade structures rather than in the external sector. A rescheduling of some foreign debts would improve the financial position significantly; a Paris Club Meeting should be a first step in this direction. Preparation of good investment projects in the public sector will improve performance; this is linked to institutional strengthening in most public entities.

4.31   It is extremely important for the country to complete the adjustment it has begun. Many difficult aspects already have been carried out. What is missing is better monetary programming and action on structural problems in the financial, public enterprises, taxes and customs areas. It is also important to receive enough backing from abroad. If the policies are implemented soon, financial stability and sustainable growth would be outcomes in the short term. But if these changes are not made, not only will no progress be made but the important gains made would be lost and old problems revived. Sufficient international cooperation will be indispensable. Too little will lead to increasing arrears. Too much will discriminate against exports by reducing the real exchange rate, and will make the country unnecessarily dependent on foreign savings.

REFERENCES

Baer, W. and Birch, M.: "International Economic Relations of a Small Country: The Case of Paraguay," *Economic Development and Cultural Change* (April 1987).

Baer, W. and Birch, M.: "From Inward to Outward Oriented Growth: Paraguay in the 1980s," *Journal of Interamerican Studies and World Affairs* (Fall 1986).

Baer, W. and Breuer, L.: "Expansion of the Economic Frontier: Paraguayan Growth in the 1970s," *World Development* (Summer 1984).

Birch, M.: "Pendulum Politics: Paraguayan Economic Diplomacy, 1940-1975," (Undated mimeo) Galleous.

Canese, R.: La Problematica de Itaipu (Asuncion, Paraguay Editorial Base-ECTA, 1990).

Ministerio de Obras Publicas y Comunicaciones: *Plan Nacional de Transporte 1987-1991* (Asuncion, Paraguay, Marzo 1987).

Pincus, J.: *The Economy of Paraguay* (New York, F. Praeger, 1968).

Raine, P.: *Paraguay* (New Brunswick, N.J., Scarecrow Press, 1956).

Rivarola, D. M., Heisecke, G.(eds.): *Poblacion, Urbanizacion y Recursos Humanos en el Paraguay* (Asuncion, Centro de Estudios Sociologicos, 1970).

Ugarte Centurion, D.: *Evolucion Historica de la Economia Paraguaya* (Asuncion, Paraguay, Editorial Graphis, 1983).

Warren, H.: Paraguay, An Informal History (November, The University of Oklahoma Press, 1949).

World Bank, "Paraguay, Agriculture Strategy Paper," June 1990a.

World Bank, "Environmental Issues Paper," June 1990a.

World Bank, "Paraguay, Country Economic Memorandum," January 1988.

World Bank, "Paraguay, Agriculture Sector Study," June 1984a.

World Bank, "Paraguay, Country Economic Memorandum," February 1984a.

World Bank, "Paraguay, Economic Memorandum," September 1981.

World Bank, "Paraguay, Regional Development in Eastern Paraguay," June 1978.

World Bank, "Paraguay, Current Economic Position and Prospects," July 1971.

World Bank, "Paraguay, Economic Position and Prospects," December 1965.

World Bank, "Paraguay, The Economy of Paraguay," March 1959.

World Bank, "Paraguay, Current Economic Position and Prospects," February 1954.

ANNEX I
Page 1 of 7

HISTORICAL BACKGROUND [1]

1.   Since colonial times, Paraguay has been continuously haunted by the threats posed by its mediterraneity. A thousand miles from the sea and hemmed in by far larger and more powerful neighbors, it has become one of the least developed countries in South America. Asuncion, its capital, was founded as a way station for explorers looking for the wealth of the Incas (Raine 1956, p. 6) and was originally at the heart of Spanish civilization in the southeastern part of the continent; in fact Paraguayans founded Argentina and settled Uruguay (Ibid., p. 3). For centuries afterwards its people were virtually isolated from the outside world, with its intellectuals far less successful in adopting European customs. For example, Paraguayans are the only people in Latin America for whom two languages, Spanish and Guarani (the aboriginal tongue), are of equal importance. Perhaps because of its isolation, Paraguay is the country in Latin America that can best claim to having developed a culture of its own (Ibid., p. 9).

Colonial Times--Blueprint for Economic and Political Insularity

2.   Paraguay's trade suffered from its mediterraneity as early as in the sixteenth century. Spain's policies forced all its colonial trade with Europe to take place through Spain. In the case of the Rio de la Plata region (including Paraguay), this trade had to go through Lima and Panama at exceedingly high costs for the region--so high that ports such as Buenos Aires remained inactive for centuries (Ugarte 1983, p. 77). Early in the seventeenth century, Asuncion was already dropping to secondary importance, a center of a neglected and isolated region. In 1617 the Council of the Indies split the Government of the Rio de la Plata into two: Buenos Aires and Guaira (which included Asuncion). After the division, the gap between the two regions widened further. Asuncion developed at a much slower pace than Buenos Aires. (Raine 1956, p. 44). Paraguay's economy remained at little more than subsistence level for the entire colonial period, with most commerce within the country carried out on a barter basis (Ibid., p. 329).

3.   The establishment and growth of Jesuit missions are often considered as powerful factors determining Paraguay's past and present political and institutional weaknesses (Ibid., p. 45). The origin of these missions stemmed from the inability of the colonial armies to subjugate local Indians. In 1608, Phillip III, King of Spain, had ordered that the Indians of Paraguay must be subdued "by the sword of the word." The Governor of the Rio de la Plata gave the Jesuits that task while granting them full power in all settlements or "reducciones" that they should start and maintain (Warren 1949, p. 85). The Jesuits were given many commercial privileges by the King, benefits that did not apply to the secular province. The Jesuits proved to be good businessmen who profited from their special privileges. The other settlers, on the other hand, had their commercial activities hampered by restrictions, taxes, labor shortages, and lack of indispensable river transportation facilities. In fact, between the early 1600s and 1767 (when the King of Spain issued a decree expelling the Jesuits from all its domains), the Paraguayan province could best be described as two independent states, one a lay municipal government; the other a theocracy run by the Jesuits (Raine 1956, p. 50).

---

[1]   This section borrows heavily from Raine (1956).

4.	The carefully planned system of life in the religious settlements made them essentially self-sufficient socially and economically. The Jesuit Fathers trained the Indians in a multiplicity of trades, teaching them to become weavers, craftsmen in iron, silver, and gold, painters, and printers. Their printing shops compared favorably with those in Europe (Ibid., p. 57). A most important task was the collection of the "yerba maté," a Paraguayan tea highly prized outside Paraguay and hence exported by the Jesuit settlements. Their settlements occupied the finest agricultural sections of the country and monopolized trade with the Argentine provinces. In fact, their monopoly was so thorough that they often threatened Asuncion with hunger to keep their privileges (Ibid., p. 61).

5.	For obvious reasons, the "secular" Paraguay resented the privileges granted to the Jesuits. Between the middle of the seventeenth century until independence in the early nineteenth century their anger was reflected in almost continuous uprising against their Authorities. The "Revolt of the Comuneros" (spanning between 1730 and 1735) was especially important. Although it pitted the local Creoles (born in America from Spanish origin) against the Spanish-designated Authorities, the true problem was the differential treatment they and the Jesuit missions received. In the years of the revolt, Spain could not control Paraguay and the communes (local people) became the supreme authority. The Viceroy of Peru tried to subdue Paraguay by cutting it off completely from the rest of the Continent and having all its goods embargoed. As this was not effective soon enough, an army was brought from Buenos Aires to put down the insurrection. This was not difficult, since in the last years of the rebellion the province was weakened and almost bankrupt--although the Jesuit missions continued to prosper. Matters deteriorated further for the secular province in 1739 when a decree from Lima forced all Paraguayan merchandise passing through the port city of Santa Fe to pay a ruinous tax, which even Buenos Aires complained about to the King of Spain, arguing that it as well as Santa Fe lived largely off Paraguayan commerce (Ibid., p. 68). After 1767, when the Jesuits were expelled, a long period of economical and political decline set in for the whole colony.

6.	Though some writers view the Jesuit experiment as a "vanished Arcadia," others contend that it was the origin of the weaknesses that persists in the Paraguayan nation even today: It failed to instill any element of self-reliance and judgement in the indigenous population[2] thus settling a fertile ground for an easy acquiescence to the dictators that later ruled Paraguay (Ibid., p. 60).

---

[2]	Two Fathers residing in the missions were omnipotent rulers despite some token communal representation (Alcaldes and Cabildantes) (Raine 1956, p. 54). Still, the Indians were happy with the regime because of the absolute equality under which they lived. They liked it so much that when a more traditional management system and social organization was imposed after the Jesuits' expulsion, the rich and populous settlements vanished in less than 20 years. (Ibid. p. 59).

Dictatorial Regimes and Unsettled Institutional Relations (1811-1870)

7. During the revolution of Spanish America, Paraguay declared its independence in July 1811, at the instigation of Dr. Jose Gaspar Rodriguez Francia. Buenos Aires had claimed Paraguay as one of its provinces, thus rejected the declaration of independence by Paraguay. Instead, it sent an emissary to negotiate a pact of union. In the subsequent agreement Buenos Aires accepted among others that: its monopoly on Paraguayan tobacco would be abolished; the tax on yerba maté then collected by Buenos Aires was going to be paid in Asuncion, although Buenos Aires might still levy a small tax on Paraguayan products in case of emergency; the Paraguay and Buenos Aires "provinces" were independent, although united in a federation and indissoluble alliance (Raine 1956, p. 86). Soon thereafter, Buenos Aires established a double duty on imported tobacco (Ibid., p. 87), and in 1813 placed heavy duties on all Paraguayan products. The tax on tobacco all but ruined the flourishing export trade on this commodity; soon afterwards, Paraguay was isolated from the world beyond its borders.

8. In 1814 Paraguay chose Dr. Francia supreme ruler for three years and then dictator for life in 1816; he died in 1840. About him, Raine (1956, p. 79) writes "seldom can be said that one man alone made a nation but that is true of Francia." He tried and succeeded in: creating an egalitarian state through virtual liquidation of the Spanish and Creole elements in favor of the Mestizo and Indian populations; destroying the power of the church; and establishing complete self-sufficiency through the almost complete isolation of the country.

9. The system of state ownership was imposed and extended to all parts of the country; roads were built and agriculture production encouraged. Francia's acts to isolate Paraguay can hardly be called voluntary; according to Raine (1956, p. 104), they were forced on him by events in neighboring countries. However, even after Buenos Aires allowed resumption of trade, Francia permitted it only under rigorous regulatory conditions. Of course, prices on imported articles were enormously high while exportable products were worth almost nothing (a man's imported hat for example was worth 60 horses) (Ibid., p. 103). In the meantime, the valuable export trade for which Paraguay had been famous was cut off and other nations quickly began to take its place in world trade.

10. Some social conditions deteriorated steadily under Francia. Although primary education was compulsory and free, with few illiterates to be found, secondary education was prohibited. Importation of books was allowed but few arrived and the President personally censored them all. Only he in all the country received newspapers from abroad (Ibid., pp. 103-4). Francia rose to power as the champion of the underprivileged and virtually wiped out the European-Hispanic element that had exploited the Indians. Like the Jesuits before him, he did not betray his constituencies, but once in power he did nothing to raise their economic independence and political maturity.

11. Soon after Dr Francia's death on September 20, 1840, Carlos Antonio Lopez took control, initially as "first Consul" of the republic (March 13, 1841), and then commanding all the absolute power and prerogatives of Dr. Francia (in 1844). He also ruled until his death on September 10, 1862. In the social area, secondary schools were founded, ports were opened, and foreigners admitted; three short highways and a railway were built, the latter, one of the first in South America (Warren 1949, p. 181). Except for ending Paraguay's isolation from the outside world, Lopez followed Francia's policies. He continued to add to national property holdings at

the expense of remaining private owners, extended the state monopoly of trade to tobacco, and kept a tight reign on foreign trade (Raine 1956, p. 156). Lopez' as well as Francia's economic and political system was similar to that of the expelled Jesuits, with the Government assuming the role the Jesuits had previously played (Ibid., p. 122).

12. In the international arena, Lopez had three main goals: gain general recognition of Paraguay's independence; settle pending boundary disputes with Brazil and the United Provinces of Argentina; and open the country to foreign trade. In 1852, the Argentine Confederation recognized Paraguay's independence; in exchange for a clear recognition of Paraguayan claims to the Chaco, Paraguay admitted Argentine's claims to the rich Misiones region (Ibid., p. 127). Also, Argentina opened the Parana River to Paraguayan commerce; thus, for the first time since the founding of the Republic, Paraguay's agricultural products could be freely exchanged for manufactures from other parts of the world. However, success was not so clear in the relations with the then Brazilian Empire. Long-standing boundary disputes were not settled, and, as a corollary, Brazilian navigation of the Parana and Paraguay Rivers over Paraguayan territory was not allowed on several occasions.

13. After Carlos Antonio Lopez death, he was succeeded by his eldest son, Mariscal Francisco Solano Lopez, who again followed the absolute policies inherited from the two previous governments (Ibid., p. 156). Although Mariscal Lopez was especially interested in developing communications (telegraph lines were installed and roads built), most public money was allocated to military expenditures and his rule is chiefly remembered by the disastrous War of the Triple Alliance. Following Paraguay's separate clashes with Brazil and Argentina, a secret Triple Alliance Treaty was signed on May 1, 1865 by Argentina, Brazil, and Uruguay. The treaty specified military arrangements for actions against Paraguay, defined future boundaries in a way to award Argentina and Brazil all previous territorial claims against Paraguay, and stated that Paraguay would have to bear all costs of any ensuing war. The treaty also provided for free navigation of the Parana and Paraguay Rivers (a vital concession for Brazil) and committed the three Allies to respect the independence and sovereignty of Paraguay.

14. Paraguay stood no military chance against its adversaries. The length of the war, the endurance of Paraguay, and the scope of the destruction however were astonishing. In 1870, five years after the beginning of the war, only a few starving Paraguayans remained on a land stripped of crops, cattle, schools, and all signs of industry. Only the fecundity of the Paraguayan's soil saved the survivors from further misery and death. Raine estimates the Paraguayan population at the beginning of the war at 800,000[3]; by the end it was reduced to less then 250,000, of whom about 150,000 were women, 80,000 children, and only 14,000 men (mostly Paraguayan fighting against their own government) (Raine 1956, p. 195). Mariscal Lopez died in the last battle of the war. As with the strong Jesuit regime, the dictatorships ruling Paraguay between 1814 and 1870 brought initial success and then ruin to the economy and the people that took decades to reverse.

---

[3] This may be an overestimate according to Mendoza (see Table 1.1).

Table 1.1: PARAGUAY: POPULATION 1865-1950
(Selected Years)

Years	Population (thousands)	
	A. Raines	B. Mendoza
1865	800	600
1872	231	231
1887	329	328
1899	635	430
1909	650	541
1919	800	683
1938	950	1062
1945	1100	1247
1950	1405	1397

Source: Raines 1956, p. 295; Mendoza, in Rivarola and Heisecke (eds.) 1970, p. 17 & p. 21

Institutional Instability Until the Chaco War (1870-1935)

15.	While foreign troops occupied the country, a new Paraguayan Constitution was approved in 1870, remaining in effect until 1940. Paraguay signed a treaty with Brazil in January 1872 ceding all territory claimed by Brazil. In December 1873 a treaty was signed with Uruguay that provided free navigation privileges and recognized a Paraguayan debt for Uruguay's war costs. In February 1876 a treaty was signed with Argentina whereby Paraguay gave up the Misiones region and the Chaco between the Bermejo and Pilcomayo Rivers. The Chaco between the Pilcomayo and the Verde River was to be submitted to an arbitral decision by the President of the United States.[4] As a result of these treaties, on June 1876, eleven years after the outbreak of hostilities, the Brazilian occupation forces completed their withdrawal from Paraguay. Uruguay renounced its indemnization awards by a treaty signed in 1883.[5]

16.	Despite the improvement achieved by the treaties signed in the 1870s, the size of the war losses left Paraguay politically and economically unstable for decades. Although the Constitution provided for four-year Presidential terms, normal political turnovers were few, and revolutions, often bloody, accounted for most of the thirty-one presidential changes between 1870 and 1928 (just before the Chaco War). Economic recovery was severely constrained.

---

[4]	On November 12, 1878, President Rutherford B. Hayes issued its arbitral award entirely in favor of Paraguay.

[5]	Brazil and Argentina also renounced their war indemnifications but almost 60 years later. Argentina in August 1942 and Brazil in May 1943 (Birch, n.d. p.11).

17. The public sector had no revenue sources and the Government could not obtain new foreign loans since it was unable to pay the interest on the old loans. The Government then resorted to land sales (most land in Paraguay became Government property under the Francia and the two Lopez regimes). The Land Acts of 1885 paved the way for the sale of these properties. These sales created an ephemeral economic boom; land prices increased ten times in eastern Paraguay and four times in the Chaco--but the Government kept selling property at fixed prices. Also because Paraguayans had little money, property fell into foreign hands, with phenomenal profits being made in the process (Raine 1956, p. 214). Another law suspending further sale of public land was passed in 1904; however, as early as 1891, almost no land was said to remain in State hands (Ibid., p. 212).[6]

18. In 1904, after 30 years of Colorado Party rule, the Liberal Party took control for nearly another 30 years (Raine 1956, p. 216).[7] Instability was rampant during the Colorado years, with only three presidents completing their terms peacefully. No improvement occurred during Liberal rule, with only two Presidents completing the four-year term. In the first eight years of Liberal Party rule, 10 presidents followed in quick succession with four in 1911 alone. The Homestead Law was enacted during President E. Schaerer's term (1912-1916), providing free land to landless Paraguayans who complied with minimal stipulations (Raine 1956, p. 219).

19. It was not until Eligio Ayala's constitutional term (1924-28), more than 50 years after the War of the Triple Alliance, that Paraguay's finances were stabilized. President Ayala reduced the 100 million peso deficit, stabilized the exchange rate, increased revenue collections, reduced foreign debt by nearly 20 percent, and retired 57 million of paper issue (maintaining a metal backing of 45 to 55 percent). In his time, as well as in his successor's (E. Guggieri), Paraguay rose to third in school attendance throughout Latin America (Ibid., p. 221). Land distribution to landless Paraguayans continued (albeit at a slow pace), and foreign debt payments were made on time. But war loomed again in the horizon, this time with Bolivia, and Paraguay could not make faster economic progress in those years due to the cost of the military buildup.

20. Responsibility for the start of the war has not been fully determined; however, armed skirmishes started in February 1927 over Bolivia's claims to the Chaco. In June 1932, Bolivian forces attacked and took Paraguay's outpost, Carlos Antonio Lopez and Paraguay officially proclaimed a state of war on May 10, 1933. As in the War of the Triple Alliance, the military odds were against Paraguay. Raine (1956) estimated Bolivia had about 10 times more manpower and equipment and 3 times more economic resources. But the terrain was more familiar to Paraguayans, who prevailed, crossing the whole Chaco region and reaching the foothills of the Andes in 1935. A Bolivian counterattack had just begun on June 14, 1935 when a general cease-fire ended the war. About 25,000 Paraguayans died and many more were wounded, crippled, or sick with malaria and other debilitating diseases. This was a cost an underpopulated country like Paraguay could ill afford and it was again near collapse. A nationalistic revolution took place in February 1936 and Eusebio Ayala, the constitutional President, lost power, ending three decades of Liberal Party rule.

---

[6] This may be an exaggerated claim, since in 1946, although just 3 percent of agriculture land in the eastern region was in public hands, 40 percent of the Chaco belonged to the public sector (Ugarte 1983, p. 138).

[7] The Colorado party was founded in 1874, the Liberal Party in 1887 (Warren 1949, p. 264).

Cycle of Strong Regimes and Instability (1936-1954)

21.   With the deprivations following the Chaco War, the army and navy revolted against the President, and in February 1936, Colonel Rafael Franco, a Chaco War hero, came to power. His socialist agenda ushered in a new political movement, the "Febreristas," which would have importance for the next two decades. However, in a country where capital was mostly owned by foreigners (primarily Argentines), Colonel Franco was unable to implement his reform program; and was forced to resign in August 1937. In July 1938, the Treaty of Peace, Friendship, and Boundaries was signed with Bolivia which settled war claims and established the borders that now prevail, with most of the Chaco going to Paraguay. Another Chaco War hero, General J. Felix Estigarribia, was elected President in 1939, with the largest number of votes ever recorded by any previous candidate. This was an auspicious beginning, but there was little time for progress, since Estigarribia died tragically in a plane crash in September 1940. Before his death he was able to have a new Constitution approved in 1940. The document gave more power to the Executive Authority, with elections held every five years; however, it failed to deal with the institutional instability it intended to cure. His most important achievement is generally considered the enactment of the Agrarian Law of 1940, which was designed to provide the basis for the revival of the country's agriculture through improved production methods, increased acreage, and wider distribution of land ownership.

22.   Colonel Higinio Morinigo, Estigarribia's War Minister, inherited the Presidency and assumed the quasi-dictatorial powers of his predecessor. During his term, the Agriculture Law was used to create national agriculture colonies, as opposed to land given to foreign immigrants in previous regimes. Morinigo also inherited an explosive labor situation he tried to contain by: drafting a labor code (not approved); adopting minimum wage policies in 1942; and setting up a social security system in January 1944. In 1946, Morinigo gave up some of his authority and began to work actively with the political parties. In February 1947, he reshuffled his cabinet and for the first time since 1904 granted full power to the Colorado Party. Despite this, a revolution broke out in March 1947 and lasted several months.

23.   Morinigo was finally deposed in June 1948, ending what Raine (1956, p. 266), calls "eight years [that] were among the most fruitful in the growth of modern Paraguay." In international relations Morinigo remained on good terms with Argentina and greatly improved economic relations with Brazil, to escape from riverway transport monopolies that inhibited the country's development at the time.

24.   As in the case of previous strong regimes, substantial instability followed Morinigo's downfall. Five Presidents held office between June 1948 and September 1949. In August 1954, General Stroessner took control and governed for the next 35 years. He was ousted in February 1989 by General Rodriguez, who was elected President in May of that year in a national election where political groups were free to participate. A more detailed analysis of the economic aspects of Paraguayan's history after the Chaco War is discussed next.

ANNEX II

SUMMARY OF SCENARIOS LOW, MEDIUM AND HIGH

8:38 AM
26-Sep-91

Paraguay Projections Model

	1988	1989	1990	1991	1992	1993	1994	1995
Population (Thousands)	4039.2	4157.3	4278.9	4404.0	4532.8	4665.4	4801.8	4942.2
Parallel Exchange Rate (G/$, base year)	136.0	136.0	136.0	136.0	136.0	136.0	136.0	136.0
Investment Financing (Bln. 1982 Guaranies)								
Total Savings	201.4	204.1	204.2	212.9	216.7	218.3	220.0	224.2
Private Savings	143.4	201.2	142.7	143.6	153.5	144.1	139.1	137.6
Financial Savings	-5.8	36.1	0.5	12.5	-4.0	-2.6	-2.6	0.2
Non-financial Savings	149.2	165.1	142.2	131.1	157.5	146.7	141.6	137.4
Public Savings	-2.2	26.7	52.2	50.9	27.1	18.3	7.9	-0.4
Foreign Savings	44.6	-47.6	-18.0	7.8	14.8	18.3	25.8	30.1
Inflation Tax	15.6	23.7	27.2	10.5	21.3	37.5	47.1	56.8
Total Investment	201.4	204.1	204.2	212.9	216.7	218.3	220.0	224.2
Private Investment	137.1	142.6	158.0	166.6	169.8	171.2	173.4	176.3
Public Investment	64.3	61.5	46.2	46.4	47.0	47.1	46.5	47.9
Investment Financing (% of GDP)								
Total Savings	23.1	22.1	21.9	22.7	21.9	21.1	20.5	20.3
Private Savings	16.5	21.8	15.3	15.3	15.5	13.9	13.0	12.5
Financial Savings	-0.7	3.9	0.1	1.3	-0.4	-0.3	-0.2	0.0
Non-financial Savings	17.1	17.9	15.2	14.0	15.9	14.2	13.2	12.5
Public Savings	-0.3	2.9	5.6	5.4	2.7	1.8	0.7	0.0
Foreign Savings	5.1	-5.2	-1.9	0.8	1.5	1.8	2.4	2.7
Inflation Tax	1.8	2.6	2.9	1.1	2.2	3.6	4.4	5.2
Total Investment	23.1	22.1	21.9	22.7	21.9	21.1	20.5	20.3
Private Investment	15.8	15.4	16.9	17.8	17.1	16.5	16.2	16.0
Public Investment	7.4	6.7	4.9	4.9	4.7	4.5	4.3	4.3
Memo items:								
Average Propensity to Save	17.8	25.4	18.0	17.5	17.5	15.8	14.7	14.2
Marginal Propensity to Save	90.9	-397.5	-3015.8	3.1	17.6	-26.4	-15.8	-5.6

Low Case

National Accounts (Bln. 1982 Guaranies)	1988	1989	1990	1991	1992	1993	1994	1995
Total Consumption	728.1	703.9	721.5	737.0	790.5	839.2	879.1	912.0
Private Consumption	686.1	647.8	660.9	675.1	720.5	765.8	803.2	832.6
Public Consumption	42.0	56.2	60.5	61.9	70.1	73.4	75.9	79.4
Total Fixed Investment	167.0	184.8	203.4	212.1	215.9	217.5	219.1	223.3
Private Investment	113.6	129.1	157.4	165.9	169.1	170.5	172.8	175.6
Public Investment	53.3	55.7	46.0	46.2	46.8	46.9	46.3	47.7
Exports	157.1	196.5	229.8	234.6	239.3	244.5	242.5	240.4
Imports	201.9	185.7	227.3	228.6	243.4	259.6	265.8	269.4
GDP	850.2	899.5	927.3	955.1	1002.3	1041.5	1075.0	1106.3
Terms-of-trade Adjustment	20.2	23.7	6.1	-17.4	-12.3	-4.8	-4.1	-3.5
GDY	870.4	923.2	933.4	937.7	990.0	1036.7	1070.9	1102.8
Transfers	5.4	3.4	4.3	4.2	4.2	4.3	4.2	4.2
Net Factor Income	-11.7	-6.8	0.5	1.7	1.3	1.3	1.3	0.9
GNY	864.1	919.8	938.2	943.6	995.6	1042.3	1076.5	1108.0
GNP	843.9	896.2	932.1	961.0	1007.8	1047.1	1080.6	1111.5
P GDY Deflator (1982=1) a/	3.81	5.12	6.98	8.03	10.44	15.66	25.05	42.59
Inflation	23.10%	34.35%	36.28%	15.00%	30.00%	50.00%	60.00%	70.00%
P Investment Deflator (1982=1)	4.60	5.66	7.01	8.06	10.48	15.72	25.14	42.75
P Inv. Deflator / P GDY Deflator	1.21	1.10	1.00	1.00	1.00	1.00	1.00	1.00
P Consumption Deflator	3.69	4.68	6.89	8.07	10.50	15.74	25.17	42.70
ICOR	3.17	3.39	6.64	7.31	4.50	5.50	6.50	7.00
MUV	1.34	1.35	1.43	1.57	1.58	1.58	1.61	1.67
GDP Growth	6.4	5.8	3.1	3.0	4.9	3.9	3.2	2.9
Private Consumption Growth	3.7	-5.6	2.0	2.1	6.7	6.3	4.9	3.6

National Accounts (Bln. Current Guaranies)	1988	1989	1990	1991	1992	1993	1994	1995
Total Consumption	2686.4	3292.9	4968.5	5950.3	8304.6	13212.9	22130.4	38945.1
Private Consumption	2531.3	3030.1	4551.5	5450.4	7568.7	12057.0	20219.8	35553.4
Public Consumption	155.1	262.7	417.0	500.0	735.9	1156.0	1910.6	3391.7
Total Fixed Investment	768.2	1045.6	1425.4	1709.6	2262.0	3417.5	5510.0	9546.5
Private Investment	522.8	730.5	1103.0	1337.3	1771.9	2680.0	4344.7	7506.5
Public Investment	245.4	315.1	322.4	372.2	490.2	737.5	1165.3	2040.0
Exports	974.1	1723.2	2219.5	2490.1	3214.0	4785.3	7085.5	11127.1
Imports	1109.5	1453.3	2139.0	2620.9	3447.2	5184.1	7898.0	12652.8
GDP	3319.1	4608.4	6474.4	7529.1	10333.5	16231.7	26827.9	46965.8
External Transfers	29.6	26.8	40.5	47.9	59.8	85.1	126.1	198.8
Net Factor Income	-64.2	-52.9	4.8	19.9	19.0	26.9	39.5	44.6
GNP	3254.9	4555.5	6479.3	7549.0	10352.5	16258.6	26867.3	47010.4
GDP US$	3951.3	4114.6	5263.8	5661.0	6276.6	6924.8	7864.4	9056.5
GNP US$	3874.9	4067.4	5267.7	5675.9	6288.1	6936.3	7876.0	9065.1
Per capita GDP (US$)	978.3	989.7	1230.2	1285.4	1384.7	1484.3	1637.8	1832.5
Per capita GNP (US$)	959.3	978.4	1231.1	1288.8	1387.3	1486.8	1640.2	1834.2

a/ Current GDP/Constant GDY; 1+inflation

Low Case

The Balance of Payments (Millions US$)	1988	1989	1990	1991	1992	1993	1994	1995	
Exports	1159.6	1538.6	1804.5	1872.3	1952.2	2041.5	2077.1	2145.7	94
Goods	871.0	1166.5	1392.3	1449.2	1510.7	1581.0	1609.0	1662.3	95
Non-factor Services	288.6	372.1	412.2	423.1	441.5	460.5	468.1	483.4	96
									97
Imports	1320.8	1297.6	1739.0	1970.6	2093.9	2211.6	2315.2	2439.9	98
Goods	1030.1	1001.3	1353.6	1583.7	1674.8	1758.6	1836.4	1930.5	99
Non-factor Services	290.7	296.3	385.4	386.9	419.0	453.0	478.9	509.4	100
									101
Resource Balance	-161.2	241.0	65.5	-98.3	-141.6	-170.1	-238.2	-294.2	102
									103
Factor Services (net)	-76.4	-47.2	3.9	14.9	11.6	11.5	11.6	8.6	104
Interest Payments	-130.3	-130.9	-90.0	-108.2	-112.4	-112.8	-114.1	-121.9	105
Others	53.9	83.7	93.9	123.1	124.0	124.2	125.7	130.5	106
									107
Transfers	35.2	23.9	32.9	36.0	36.3	36.3	37.0	38.3	108
									109
Current Account	-202.4	217.7	102.3	-47.4	-93.8	-122.3	-189.6	-247.3	110
									111
Capital Account	-1.5	-80.2	-14.6	217.4	-66.2	-47.7	4.6	17.3	112
									113
Direct Foreign Investment	0.0	0.0	73.1	30.0	33.3	36.6	41.0	46.7	114
									115
Net MLT Flows	-83.4	165.1	-576.1	-62.5	-44.5	-52.2	-56.1	-18.0	116
Disbursements	141.8	628.8	80.0	140.9	141.2	127.0	101.1	147.8	117
Amortizations	225.2	463.7	656.1	203.4	185.7	179.2	157.3	165.8	118
									119
Other Capital Flows	-99.9	-4.6	352.0	70.0	0.0	0.0	0.0	0.0	120
									121
Other	181.8	-240.7	136.4	150.0	-90.3	-90.3	-92.0	-95.4	122
									123
Financing Gap	0.0	0.0	0.0	29.8	35.3	58.3	111.8	83.9	124
									125
Overall BOP Surplus	-203.9	137.5	87.7	170.0	-160.0	-170.0	-185.0	-230.0	126
Change in Net International Reserves	-143.8	136.7	245.9	300.0	20.0	20.0	20.0	20.0	127
Arrears	60.1	-0.8	158.2	130.0	180.0	190.0	205.0	250.0	128
Amortizations	44.4	13.9	122.2	100.0	150.0	150.0	155.0	180.0	129
Interest	15.7	-14.7	36.0	30.0	30.0	40.0	50.0	70.0	130
									131
Memo Items:									132
Nominal Exchange Rate (Average)	840.0	1120.0	1230.0	1330.0	1646.3	2344.0	3411.3	5185.8	133
Real Exchange Rate	220.3	218.6	176.3	165.6	157.7	149.7	136.2	121.8	134
Implicit Exchange Rate on Imports		218.6	176.2	165.6	170.3	167.7	163.4	158.3	135
QR Policy Variable		1	1	1	1.08	1.12	1.2	1.3	136
Real Eff. Exch. Rate (1982=1)	1.4411	1.5276	1.3477	1.4281	1.3567	1.2753	1.1860	1.1030	137
Amortization (DRS)	236.7								138
BOP Interest - DRS Interest	-77.5	-.39	-114.5	-144.5	-174.5	-214.5	-264.5	-334.5	139
Terms of Trade (1982=100)	116.1	113.6	97.9	92.5	94.9	98.3	98.5	98.7	140
Gap plus Net MLT	-83.4	165.1	-576.1	-32.6	-9.2	6.1	55.6	65.9	141
Net Binationals									142
Stock of Int'l. Reserves	291.2	427.9	673.8	973.8	993.8	1013.8	1033.8	1053.8	143
Low Case									144

Public Sector (% of GDP)	1988	1989	1990	1991	1992	1993	1994	1995	
Current Revenues	14.3	17.3	18.9	19.0	17.7	17.7	17.6	17.5	308
Taxes	7.0	8.1	9.2	9.3	8.0	8.0	7.9	7.8	309
Imports	1.1	1.6	2.3	2.6	1.4	1.3	1.2	1.1	310
Exports	0.0	0.8	0.5	0.0	0.0	0.0	0.0	0.0	311
Direct	2.2	1.8	1.5	1.7	1.7	1.7	1.7	1.7	312
Indirect	3.7	3.9	5.0	5.0	5.0	5.0	5.0	5.0	313
Non-tax Revenues	3.8	6.1	5.4	5.4	5.4	5.4	5.4	5.4	314
V.A. on Public Enterprises	3.4	3.1	4.3	4.3	4.3	4.3	4.3	4.3	315
									316
Current Expenditures	10.0	13.8	13.3	13.5	14.0	13.9	13.8	14.0	317
Government Salaries	2.9	3.8	4.4	4.6	5.0	5.0	5.0	5.1	318
Public Enterprise Salaries	1.5	1.5	1.5	1.5	1.6	1.6	1.6	1.6	319
Social Security & PE Tax	0.6	0.9	0.9	0.9	0.9	0.9	0.9	0.9	320
Other Government	1.7	1.9	2.0	2.0	2.1	2.1	2.1	2.1	321
Transfers SP	1.9	2.4	2.3	2.3	2.4	2.4	2.4	2.4	322
Interest	1.4	3.2	2.2	2.2	1.9	1.9	1.8	1.9	323
External	1.2	3.1	2.1	2.1	2.0	1.9	1.8	1.8	324
Domestic	0.2	0.1	0.2	0.1	0.0	0.0	0.0	0.1	325
									326
Exchange Rate Adjustments	4.6	0.5	0.0	0.0	1.0	2.0	3.0	3.5	327
									328
Current Savings	-0.3	3.0	5.6	5.4	2.7	1.8	0.7	0.0	329
									330
Capital Expenditures	7.4	6.8	5.0	4.9	4.7	4.5	4.3	4.3	331
Direct Investment	5.9	5.0	4.4	4.9	4.7	4.5	4.3	4.3	332
Indirect Investment	1.5	1.8	0.5	0.0	0.0	0.0	0.0	0.0	333
									334
Deficit	7.6	3.9	-0.7	-0.5	2.0	2.8	3.6	4.4	335
External Financing	2.3	1.5	1.5	2.0	3.1	2.3	2.6	2.8	336
Disbursements	2.3	1.5	1.5	2.6	2.6	1.6	1.6	1.5	337
Amortizations (program.)	0.0	0.0	0.0	2.9	2.4	2.1	1.6	1.5	338
Interest Arrears	0.0	0.0	0.0	0.5	0.5	0.6	0.6	0.8	339
Amortization Arrears	0.0	0.0	0.0	1.8	2.4	2.2	2.0	2.0	340
Domestic Financing	5.3	2.4	-2.1	-2.5	-1.1	0.5	1.0	1.5	341
Banks	-0.8	-2.4	-2.2	-1.6	-0.8	0.4	0.7	1.2	342
Other	6.1	4.9	0.1	-0.8	-0.3	0.1	0.2	0.4	343
Low Case									346

Financial Sector (Bln. of Guaranies, stock)	1988	1989	1990	1991	1992	1993	1994	1995	
Total Assets	483.0	856.9	1231.3	1408.1	1767.9	2576.6	4016.6	6790.2	347
Net International Reserves	287.7	536.9	891.9	1275.9	1305.6	1345.5	1403.1	1489.1	348
Domestic Credit	368.2	407.0	440.1	232.9	497.0	1123.3	2167.1	4284.8	349
Public	9.3	-103.4	-247.7	-370.0	-453.8	-393.2	-197.4	346.9	350
Private	358.9	510.3	687.7	602.9	950.8	1516.5	2364.4	3937.9	351
Exchange Losses	100.6	13.8	0.0	0.0	103.3	324.6	804.8	1643.8	352
Other Net Assets	-273.5	-100.6	-100.6	-100.6	-138.0	-216.8	-358.4	-627.4	353
									354
Total Liabilities	483.0	856.9	1146.3	1408.1	1767.92	2576.6	4016.6	6790.2	355
									356
M1	263.1	384.3	490.7	570.6	783.1	1230.1	2033.1	3559.2	357
Currency	149.1	216.2	300.5	349.5	479.6	753.4	1245.3	2180.0	358
Demand Deposits	114.0	168.1	190.1	221.1	303.5	476.7	787.8	1379.2	359
									360
QM	185.6	403.6	586.6	768.6	915.8	1277.5	1914.5	3162.0	361
Domestic Currency	167.6	225.2	290.9	413.6	547.0	826.7	1312.7	2298.1	362
Foreign Currency	18.0	178.4	295.7	355.0	368.9	450.8	601.8	863.9	363
									364
Foreign Liabilities LT	34.3	69.0	69.0	69.0	69.0	69.0	69.0	69.0	365
									366
Memo Items:									367
Net Internat. Reserves (Millions US$)	342.5	479.2	725.1	1025.1	1045.1	1065.1	1085.1	1105.1	368
Foreign Liabilities L.P. (Millions US$)									369
Unaccounted Funds (Millions US$)	-199.8	29.1	-674.3	-166.2	-214.9	-141.2	-121.5	-133.8	370
									371
Financial Sector (Billions 1982 Guaranies, stock)									372
Total Assets	126.7	167.3	176.4	175.4	169.4	164.6	160.3	159.4	373
Net International Reserves	75.4	104.8	127.7	158.9	125.1	85.9	56.0	35.0	374
Domestic Credit	96.6	79.4	63.0	29.0	47.6	71.7	86.5	100.6	375
Public	2.4	-20.2	-35.5	-46.1	-43.5	-25.1	-7.9	8.1	376
Private	94.1	99.6	98.5	75.1	91.1	96.9	94.4	92.5	377
Exchange Losses	26.4	2.7	0.0	0.0	9.9	20.7	32.1	38.6	378
Other Net Assets	-71.7	-19.6	-14.4	-12.5	-13.2	-13.8	-14.3	-14.7	379
									380
Total Liabilities	126.7	167.3	164.2	175.4	169.4	164.6	160.3	159.4	381
									382
M1	69.0	75.0	70.3	71.1	75.0	78.6	81.2	83.6	383
Currency	39.1	42.2	43.0	43.5	46.0	48.1	49.7	51.2	384
Demand Deposits	29.9	32.8	27.2	27.5	29.1	30.4	31.4	32.4	385
									386
QM	48.7	78.8	84.0	95.7	87.7	81.6	76.4	74.2	387
Domestic Currency	44.0	44.0	41.7	51.5	52.4	52.8	52.4	54.0	388
Foreign Currency	4.7	34.8	42.4	44.2	35.3	28.8	24.0	20.3	389
									390
Foreign Liabilities LT	9.0	13.5	9.9	8.6	6.6	4.4	2.8	1.6	391
Low Case									392

Debt (Pipeline + New) (Mln. US$)	1988	1989	1990	1991	1992	1993	1994	1995	
Total Commitments	211.8	0.0	0.0	224.8	165.7	208.7	267.4	248.1	446
Refinancing	0.0	0.0	0.0	0.0	0.0	0.0	0.0	0.0	447
Other Multilateral and IDB	79.9	0.0	0.0	50.0	50.4	50.4	48.8	48.2	448
World Bank	0.0	0.0	0.0	5.0	30.0	50.0	53.4	58.0	449
Brazil (CACEX)	0.0	0.0	0.0	0.0	0.0	0.0	0.0	0.0	450
Bilateral	93.6	0.0	0.0	70.0	50.0	50.0	53.4	58.0	451
Suppliers + Export Credits	0.0	0.0	0.0	0.0	0.0	0.0	0.0	0.0	452
Financial Institutions	38.3	0.0	0.0	0.0	0.0	0.0	0.0	0.0	453
Private Non Guaranteed	0.0	0.0	0.0	0.0	0.0	0.0	0.0	0.0	454
IMF	0.0	0.0	0.0	0.0	0.0	0.0	0.0	0.0	455
Short Term	0.0	0.0	0.0	70.0	0.0	0.0	0.0	0.0	456
Unidentified Sources	0.0	0.0	0.0	29.8	35.3	58.3	111.8	83.9	457
Total Disbursements	141.8	628.8	80.0	240.7	176.6	185.3	212.9	231.7	458
Refinancing	0.0	0.0	0.0	0.0	0.0	0.0	0.0	0.0	459
Other Multilateral and IDB	31.4	18.0	15.0	24.6	33.4	43.5	39.9	50.1	460
World Bank	18.0	16.1	16.0	12.3	10.6	9.6	15.2	23.8	461
Brazil (CACEX)	0.0	426.6	22.0	0.0	0.0	0.0	0.0	0.0	462
Bilateral	40.4	125.9	22.0	70.7	76.9	60.2	46.0	65.0	463
Suppliers + Export Credits	14.8	41.2	25.0	25.6	15.2	10.1	0.0	2.6	464
Financial Institutions	37.2	0.9	2.0	7.6	5.2	3.6	0.0	6.3	465
Private Non Guaranteed	0.0	0.0	0.0	0.0	0.0	0.0	0.0	0.0	466
IMF	0.0	0.0	0.0	0.0	0.0	0.0	0.0	0.0	467
Short Term	-63.0	44.0	76.0	70.0	0.0	0.0	0.0	0.0	468
Unidentified Sources	0.0	0.0	0.0	29.8	35.3	58.3	111.8	83.9	469
Amortization Arrears	44.4	13.9	122.2	100.0	150.0	150.0	155.0	180.0	470
Total Amortizations	180.8	449.8	533.9	203.4	185.7	179.2	157.3	165.8	471
Refinancing	0.0	0.0	0.0	0.0	0.0	0.0	0.0	0.0	472
Other Multilateral and IDB	13.0	22.4	21.0	27.1	29.3	30.8	31.8	40.4	473
World Bank	42.8	37.7	41.0	38.5	38.7	38.2	37.2	32.9	474
Brazil (CACEX)	0.0	0.0	426.6	0.0	0.0	0.0	0.0	0.0	475
Bilateral	41.1	59.5	28.0	44.1	44.9	45.7	42.2	59.4	476
Suppliers + Export Credits	40.3	5.7	16.0	42.5	38.1	33.0	29.6	17.9	477
Financial Institutions	43.6	323.1	0.0	49.6	32.9	29.3	14.8	8.3	478
Private Non Guaranteed	0.0	1.3	1.3	1.6	1.8	2.1	1.7	1.8	479
IMF	0.0	0.0	0.0	0.0	0.0	0.0	0.0	0.0	480
Unidentified Sources	0.0	0.0	0.0	0.0	0.0	0.0	0.0	5.0	481
Low Case									485

Debt (Pipeline + New) (Mln. US$) cont.	1988	1989	1990	1991	1992	1993	1994	1995	
Interest Arrears	15.7	-14.7	36.0	30.0	30.0	40.0	50.0	70.0	486
Total Interest	114.6	91.4	90.0	108.2	112.4	112.8	114.1	121.9	487
Refinancing	0.0	0.0	0.0	0.0	0.0	0.0	0.0	0.0	488
Other Multilateral and IDB	20.6	20.5	22.0	21.3	21.0	21.3	21.8	22.1	489
World Bank	30.2	25.6	25.0	22.2	20.1	17.8	15.6	13.9	490
Brazil (CACEX)	0.0	14.8	0.0	0.0	0.0	0.0	0.0	0.0	491
Bilateral	17.7	9.2	10.0	16.3	17.0	17.7	17.6	17.4	492
Suppliers + Export Credits	25.0	7.0	17.0	12.6	11.2	8.7	6.2	3.9	493
Financial Institutions	19.4	2.5	1.0	10.9	7.2	4.5	2.7	1.6	494
Private Non Guaranteed	0.7	1.1	0.0	1.0	1.0	0.9	0.9	0.9	495
Interest on Arrears	0.0	9.8	14.0	22.8	28.3	32.8	36.7	41.7	496
IMF	0.0	0.0	0.0	0.0	0.0	0.0	0.0	0.0	497
Short-Term	1.0	1.0	1.0	1.0	4.3	4.1	3.9	3.9	498
Unidentified Sources	0.0	0.0	0.0	0.0	2.4	5.0	8.8	16.5	499
Other Interest									500
Accumulated Arrears	282.1	281.3	439.5	569.5	749.5	939.5	1144.5	1394.5	501
Total DOD	2354.9	2388.7	2127.5	2294.9	2465.7	2661.8	2922.4	3238.3	502
Refinancing	0.0	0.0	0.0	0.0	0.0	0.0	0.0	0.0	503
Other Multilateral and IDB	419.4	409.7	415.0	412.5	416.7	429.3	437.4	447.1	504
World Bank	358.1	324.3	320.0	293.8	265.6	237.1	215.1	206.0	505
Brazil (CACEX)	0.0	426.6	0.0	0.0	0.0	0.0	0.0	0.0	506
Bilateral	478.2	450.4	470.0	496.7	528.6	543.0	546.9	552.4	507
Suppliers + Export credits	210.3	219.3	248.0	231.1	208.2	185.3	155.7	140.4	508
Financial Institutions	629.8	269.9	283.0	241.0	213.3	187.6	172.8	170.8	509
Private Non Guaranteed	28.0	28.0	19.0	17.4	15.6	13.5	11.8	10.0	510
IMF	0.0	0.0	0.0	0.0	0.0	0.0	0.0	0.0	511
Unidentified Sources	0.0	0.0	0.0	0.0	0.0	0.0	0.0	0.0	512
Short Term	231.2	260.5	372.5	472.5	502.5	542.5	592.5	662.5	513
Interest Arrears	93.2	78.5	114.5	144.5	174.5	214.5	264.5	334.5	514
Other	138.0	182.0	258.0	328.0	328.0	328.0	328.0	328.0	515
Adjustment in Debt Stock	-2.9	-3.7	-3.7	-3.7	-3.7	-3.7	-3.7	-3.7	516
Net Disbursements:									517
Other Multilateral and IDB	18.4	-4.4	-6.0	-2.5	4.1	12.6	8.1	9.7	518
World Bank	-24.9	-21.6	-25.0	-26.2	-28.1	-28.6	-22.0	-9.1	519
Brazil (CACEX)	0.0	426.6	-426.6	0.0	0.0	0.0	0.0	0.0	520
Bilateral	-0.7	66.4	-6.0	26.7	31.9	14.5	3.8	5.5	521
Memorandum:									522
Total Debt	2352.0	2385.0	2123.8	2291.2	2462.0	2658.1	2918.7	3234.6	523
Debt Service/Exports (%)	37.3	41.2	43.3	23.6	24.5	23.6	22.9	25.1	524
Debt Service/GDP (%)	11.0	15.4	16.3	9.8	7.6	9.5	8.8	8.9	525
Interest/Exports (%)	11.2	5.0	7.0	7.4	7.3	7.5	7.9	8.9	526
Interest/GDP (%)	3.3	1.9	2.4	2.4	2.3	2.2	2.1	2.1	527
Debt/Exports (%)	227.4	173.5	142.3	153.0	164.7	176.4	195.8	215.9	528
Debt/GDP (%)	66.7	64.9	48.8	50.6	51.2	52.0	51.7	51.2	529
Low Case									531

8:23 AM
26-Sep-91

Paraguay Projections Model

	1988	1989	1990	1991	1992	1993	1994	1995	1996	1997	1998	1999	2000
Population (Thousands)	4039.2	4157.3	4278.9	4404.0	4532.8	4665.4	4801.8	4942.2	5086.8	5235.5	5388.6	5546.2	5708.4
Parallel Exchange Rate (G/$, base year)	136.0	136.0	136.0	136.0	136.0	136.0	136.0	136.0	136.0	136.0	136.0	136.0	136.0
Investment Financing (Bln. 1982 Guaranies)													
Total Savings	201.4	204.1	204.2	212.9	225.8	239.1	251.5	264.6	277.2	291.5	305.4	319.9	335.1
Private Savings	143.4	201.2	142.7	143.6	147.4	162.7	168.1	173.9	179.5	187.8	196.1	205.2	217.0
Financial Savings	-5.8	36.1	0.5	12.5	12.1	14.9	12.5	13.5	13.6	14.4	15.6	16.1	17.4
Non-financial Savings	149.2	165.1	142.2	131.1	135.3	147.8	155.6	160.4	165.9	173.4	180.5	189.1	199.6
Public Savings	-2.2	26.7	52.2	50.9	62.1	61.7	65.4	69.8	75.4	79.3	82.4	84.6	86.7
Foreign Savings	44.6	-47.6	-18.0	7.8	7.0	8.0	10.9	13.5	14.4	16.3	18.4	21.2	22.1
Inflation Tax	15.6	23.7	27.2	10.5	9.2	6.7	7.1	7.5	7.8	8.2	8.6	9.0	9.4
Total Investment	201.4	204.1	204.2	212.9	225.8	239.1	251.5	264.6	277.2	291.5	305.4	319.9	335.1
Private Investment	137.1	142.6	158.0	166.6	175.1	185.3	195.2	204.5	214.2	224.4	235.1	246.4	258.2
Public Investment	64.3	61.5	46.2	46.4	50.7	53.8	56.3	60.2	63.0	67.1	70.3	73.5	77.0
Investment Financing (% of GDP)													
Total Savings	23.1	22.1	21.9	22.7	22.9	22.9	23.0	23.1	23.1	23.2	23.2	23.2	23.3
Private Savings	16.5	21.8	15.3	15.3	15.0	15.6	15.3	15.1	15.0	15.0	14.9	14.9	15.1
Financial Savings	-0.7	3.9	0.1	1.3	1.2	1.4	1.1	1.2	1.1	1.1	1.2	1.2	1.2
Non-financial Savings	17.1	17.9	15.2	14.0	13.7	14.1	14.2	14.0	13.8	13.8	13.7	13.7	13.9
Public Savings	-0.3	2.9	5.6	5.4	6.3	5.9	6.0	6.1	6.3	6.3	6.3	6.1	6.0
Foreign Savings	5.1	-5.2	-1.9	0.8	0.7	0.8	1.0	1.2	1.2	1.3	1.4	1.5	1.5
Inflation Tax	1.8	2.6	2.9	1.1	0.9	0.6	0.7	0.7	0.7	0.7	0.7	0.7	0.7
Total Investment	23.1	22.1	21.9	22.7	22.9	22.9	23.0	23.1	23.1	23.2	23.2	23.2	23.3
Private Investment	15.8	15.4	16.9	17.8	17.8	17.7	17.8	17.8	17.8	17.9	17.9	17.9	17.9
Public Investment	7.4	6.7	4.9	4.9	5.1	5.1	5.1	5.2	5.2	5.3	5.3	5.3	5.3
Memo Items:													
Average Propensity to Save	17.8	25.4	18.0	17.5	17.2	17.9	17.7	17.5	17.3	17.3	17.3	17.3	17.5
Marginal Propensity to Save	90.9	-397.5	-3015.8	3.1	11.4	28.5	12.7	13.2	12.9	17.7	16.6	17.7	21.5

Medium Case

National Accounts (Bln. 1982 Guaranies)	1988	1989	1990	1991	1992	1993	1994	1995	1996	1997	1998	1999	2000
Total Consumption	728.1	703.9	721.5	737.0	771.8	816.9	857.5	899.5	941.8	984.9	1031.5	1079.4	1128.7
Private Consumption	686.1	647.8	660.9	675.1	704.7	743.7	779.7	816.9	854.3	892.1	933.1	975.1	1018.0
Public Consumption	42.0	56.2	60.5	61.9	67.1	73.3	77.8	82.6	87.5	92.8	98.4	104.3	110.7
Total Fixed Investment	167.0	184.8	203.4	212.1	225.0	238.2	250.6	263.7	276.1	290.5	304.3	318.7	333.9
Private Investment	113.6	129.1	157.4	165.9	174.5	184.7	194.5	203.7	213.4	223.6	234.3	245.5	257.2
Public Investment	53.3	55.7	46.0	46.2	50.5	53.6	56.1	60.0	62.7	66.9	70.0	73.3	76.7
Exports	157.1	196.5	229.8	234.6	249.8	268.7	282.4	297.2	311.2	325.7	341.3	357.1	374.0
Imports	201.9	185.7	227.3	228.6	248.1	273.1	290.2	307.8	321.6	336.0	351.6	366.3	381.3
GDP	850.2	899.5	927.3	955.1	998.4	1050.7	1100.4	1152.6	1207.5	1265.0	1325.5	1388.9	1455.3
Terms-of-trade Adjustment	20.2	23.7	6.1	-17.4	-13.0	-5.6	-5.2	-4.9	-7.0	-8.9	-10.4	-12.9	-14.9
GDY	870.4	923.2	933.4	937.7	985.5	1045.1	1095.2	1147.7	1200.5	1256.1	1315.1	1376.0	1440.4
Transfers	5.4	3.4	4.3	4.2	4.2	4.3	4.2	4.2	4.2	4.2	4.2	4.1	4.1
Net Factor Income	-11.7	-6.8	0.5	1.7	2.2	0.2	1.0	1.8	3.1	3.7	3.7	3.2	2.7
GNY	864.1	919.8	938.2	943.6	991.8	1049.6	1100.5	1153.8	1207.9	1264.0	1323.0	1383.4	1447.2
GNP	843.9	896.2	932.1	961.0	1004.8	1055.2	1105.7	1158.6	1214.8	1272.9	1333.4	1396.3	1462.1
P GDY Deflator (1982=1) a/	3.81	5.12	6.98	8.03	9.07	9.89	10.78	11.75	12.81	13.96	15.22	16.59	18.08
Inflation	23.10%	34.35%	36.28%	15.00%	13.00%	9.00%	9.00%	9.00%	9.00%	9.00%	9.00%	9.00%	9.00%
P Investment Deflator (1982=1)	4.60	5.66	7.01	8.06	9.11	9.93	10.82	11.79	12.86	14.01	15.27	16.65	18.15
P Inv. Deflator / P GDY Deflator	1.21	1.10	1.00	1.00	1.00	1.00	1.00	1.00	1.00	1.00	1.00	1.00	1.00
P Consumption Deflator	3.69	4.68	6.89	8.07	9.12	9.93	10.84	11.82	12.90	14.06	15.33	16.72	18.21
ICOR	3.17	3.39	6.64	7.31	4.90	4.30	4.80	4.80	4.80	4.80	4.80	4.80	4.80
MUV	1.34	1.35	1.43	1.57	1.58	1.58	1.61	1.67	1.73	1.80	1.86	1.93	2.00
GDP Growth	6.4	5.8	3.1	3.0	4.5	5.2	4.7	4.7	4.8	4.8	4.8	4.8	4.8
Private Consumption Growth	3.7	-5.6	2.0	2.1	4.4	5.5	4.9	4.8	4.6	4.4	4.6	4.5	4.4
National Accounts (Bln. Current Guaranies)	1988	1989	1990	1991	1992	1993	1994	1995	1996	1997	1998	1999	2000
Total Consumption	2686.4	3292.9	4968.5	5950.3	7038.9	8113.2	9293.8	10636.3	12145.3	13850.4	15815.8	18043.0	20557.5
Private Consumption	2531.3	3030.1	4551.5	5450.4	6427.3	7385.5	8450.8	9659.9	11016.7	12545.8	14306.9	16299.3	18541.8
Public Consumption	155.1	262.7	417.0	500.0	611.6	727.6	843.0	976.4	1128.6	1304.7	1508.9	1743.7	2015.6
Total Fixed Investment	768.2	1045.6	1425.4	1709.6	2048.7	2364.6	2711.4	3109.5	3549.6	4069.8	4647.4	5306.2	6058.9
Private Investment	522.8	730.5	1103.0	1337.3	1588.8	1833.0	2104.2	2402.4	2743.4	3132.8	3578.0	4086.6	4667.3
Public Investment	245.4	315.1	322.4	372.2	459.9	531.6	607.2	707.1	806.2	937.0	1069.3	1219.6	1391.5
Exports	974.1	1723.2	2219.5	2490.1	3073.3	3724.4	4278.9	4927.3	5597.2	6353.1	7233.9	8201.3	9327.3
Imports	1109.5	1453.3	2139.0	2620.9	3219.9	3866.6	4478.2	5187.8	5916.6	6738.1	7685.7	8727.7	9902.8
GDP	3319.1	4608.4	6474.4	7529.1	8941.0	10335.5	11805.8	13485.4	15375.6	17535.3	20011.3	22822.8	26040.9
External Transfers	29.6	26.8	40.5	47.9	54.7	60.3	65.4	71.2	77.2	83.6	90.8	98.3	106.6
Net Factor Income	-64.2	-52.9	4.8	19.9	28.1	2.9	16.1	31.0	57.6	74.2	81.8	76.7	69.4
GNP	3254.9	4555.5	6479.3	7549.0	8969.1	10338.4	11821.9	13516.4	15433.2	17609.5	20093.1	22899.6	26110.3
GDP US$	3951.3	4114.6	5263.8	5661.0	5930.3	6226.5	6673.1	7261.0	7919.3	8644.1	9424.2	10295.0	11231.1
GNP US$	3874.9	4067.4	5267.7	5675.9	5948.9	6228.3	6682.2	7277.7	7949.0	8660.7	9462.8	10329.6	11261.0
Per capita GDP (US$)	978.3	989.7	1230.2	1285.4	1308.3	1334.6	1389.7	1469.2	1556.9	1651.0	1748.9	1856.2	1967.5
Per capita GNP (US$)	959.3	978.4	1231.1	1288.8	1312.4	1335.0	1391.6	1472.6	1562.7	1658.0	1756.1	1862.5	1972.7

a/ Current GDP/Constant GDY; 1+Inflation

Medium Case

The Balance of Payments (Millions US$)	1988	1989	1990	1991	1992	1993	1994	1995	1996	1997	1998	1999	2000
Exports	1159.6	1538.6	1804.5	1872.3	2038.4	2243.7	2418.6	2653.0	2882.9	3131.8	3406.8	3699.5	4022.7
Goods	871.0	1166.5	1392.3	1449.2	1577.4	1737.6	1873.6	2055.4	2233.8	2426.9	2640.3	2867.4	3118.3
Non-factor Services	288.6	372.1	412.2	423.1	461.0	506.1	545.0	597.6	649.1	704.9	766.5	832.1	904.5
Imports	1320.8	1297.6	1739.0	1970.6	2135.7	2329.4	2531.3	2793.3	3047.4	3321.6	3619.5	3936.9	4270.9
Goods	1030.1	1001.3	1353.6	1583.7	1707.9	1851.6	2006.8	2208.9	2407.0	2620.8	2852.3	3101.4	3362.6
Non-factor Services	290.7	296.3	385.4	386.9	427.8	477.8	524.5	584.4	640.4	700.8	767.2	835.5	908.3
Resource Balance	-161.2	241.0	65.5	-98.3	-97.2	-85.7	-112.7	-140.2	-164.5	-189.8	-212.8	-237.4	-248.2
Factor Services (net)	-76.4	-47.2	3.9	14.9	18.6	1.8	9.1	16.7	29.7	36.6	38.5	34.6	29.9
Interest Payments	-130.3	-130.9	-90.0	-108.2	-107.1	-126.8	-123.5	-124.5	-116.2	-117.7	-124.6	-137.9	-152.4
Others	53.9	83.7	93.9	123.1	125.7	128.6	132.6	141.2	145.9	154.3	163.2	172.5	182.3
Transfers	35.2	23.9	32.9	36.0	36.3	36.3	37.0	38.3	39.8	41.2	42.7	44.3	46.0
Current Account	-202.4	217.7	102.3	-47.4	-42.3	-47.6	-66.6	-85.2	-95.1	-111.9	-131.5	-158.5	-172.3
Capital Account	-1.5	-80.2	-14.6	217.4	653.1	104.1	125.5	161.6	158.6	180.5	206.0	237.8	255.8
Direct Foreign Investment	0.0	0.0	73.1	30.0	33.3	36.6	41.0	46.7	53.3	60.8	69.4	79.1	90.3
Net MLT Flows	-83.4	165.1	-576.1	-62.5	529.7	-11.4	-3.2	20.6	-66.5	-74.1	-84.7	-96.8	-131.6
Disbursements	141.8	628.8	80.0	140.9	715.4	167.8	154.1	186.4	202.1	214.1	233.5	256.7	279.5
Amortizations	225.2	463.7	656.1	203.4	185.7	179.2	157.3	165.8	268.7	288.1	318.2	353.4	411.1
Other Capital Flows	-99.9	-4.6	352.0	70.0	70.6	70.6	71.8	74.5	77.2	80.1	83.1	86.1	89.3
Other	181.8	-240.7	136.4	150.0	0.0	0.0	0.0	0.0	0.0	0.0	0.0	0.0	0.0
Financing Gap	0.0	0.0	0.0	29.8	19.5	8.4	15.9	19.7	94.6	113.6	138.3	169.3	207.8
Overall BOP Surplus	-203.9	137.5	87.7	170.0	610.8	56.5	58.9	76.4	63.5	68.5	74.5	79.3	83.5
Change in Net International Reserves	-143.8	136.7	245.9	300.0	41.3	56.5	58.9	76.4	63.5	68.5	74.5	79.3	83.5
Arrears	60.1	-0.8	158.2	130.0	-569.5	0.0	0.0	0.0	0.0	0.0	0.0	0.0	0.0
Amortizations	44.4	13.9	122.2	100.0	-425.0	0.0	0.0	0.0	0.0	0.0	0.0	0.0	0.0
Interest	15.7	-14.7	36.0	30.0	-144.5	0.0	0.0	0.0	0.0	0.0	0.0	0.0	0.0
Memo items:													
Nominal Exchange Rate (Average)	840.0	1120.0	1230.0	1330.0	1507.7	1659.9	1769.2	1857.2	1941.5	2028.6	2123.4	2216.9	2318.6
Real Exchange Rate	220.3	218.6	176.2	165.6	166.2	167.8	164.1	158.1	151.6	145.3	139.5	133.7	128.3
Implicit Exchange Rate on Imports		218.6	176.2	165.6	166.2	167.8	164.1	158.1	151.6	145.3	139.5	133.7	128.3
QR Policy Variable	1.4411	1	1	1	1	1	1	1	1	1	1	1	1
Real Eff. Exch. Rate (1982=1)		1.5276	1.3477	1.4281	1.4302	1.4316	1.4316	1.4345	1.4366	1.4366	1.4366	1.4366	1.4366
Amortization (DRS)	236.7												
BOP Interest - DRS Interest	-77.5	-39	-114.5	-144.5	0	0	0	0	0	0	0	0	0
Terms of Trade (1982=100)	116.1	113.6	97.9	92.5	94.8	98.1	98.3	98.4	97.6	97.0	96.5	95.8	95.3
Gap plus Net MLT	-83.4	165.1	-576.1	-32.6	549.2	-3.0	12.7	40.4	28.0	39.6	53.6	72.6	76.2
Net Binationals													
Stock of Int'l. Reserves	291.2	427.9	673.8	973.8	1015.1	1071.6	1130.4	1206.9	1270.4	1338.9	1413.4	1492.8	1576.3
Medium Case													

Public Sector (% of GDP)	1988	1989	1990	1991	1992	1993	1994	1995	1996	1997	1998	1999	2000
Current Revenues	14.3	17.3	18.9	19.0	19.9	20.0	20.0	20.1	20.1	20.1	20.1	20.0	20.0
Taxes	7.0	8.1	9.2	9.3	10.2	10.3	10.3	10.4	10.4	10.4	10.4	10.3	10.3
Imports	1.1	1.6	2.3	2.6	3.0	3.1	3.1	3.2	3.2	3.2	3.2	3.2	3.2
Exports	0.0	0.8	0.5	0.0	0.0	0.0	0.0	0.0	0.0	0.0	0.0	0.0	0.0
Direct	2.2	1.8	1.5	1.7	1.8	1.8	1.8	1.8	1.8	1.8	1.8	1.8	1.8
Indirect	3.7	3.9	5.0	5.0	5.4	5.4	5.4	5.4	5.4	5.4	5.4	5.4	5.4
Non-tax Revenues	3.8	6.1	5.4	5.4	5.4	5.4	5.4	5.4	5.4	5.4	5.4	5.4	5.4
V.A. on Public Enterprises	3.4	3.1	4.3	4.3	4.3	4.3	4.3	4.3	4.3	4.3	4.3	4.3	4.3
Current Expenditures	10.0	13.8	13.3	13.5	13.6	14.1	14.0	14.0	13.8	13.7	13.8	13.9	14.0
Government Salaries	2.9	3.8	4.4	4.6	4.8	5.0	5.1	5.2	5.3	5.4	5.5	5.6	5.7
Public Enterprise Salaries	1.5	1.5	1.5	1.5	1.5	1.6	1.7	1.7	1.7	1.7	1.7	1.7	1.7
Social Security & PE Tax	0.6	0.9	0.9	0.9	0.9	0.9	0.9	0.9	0.9	0.9	0.9	0.9	0.9
Other Government	1.7	1.9	2.0	2.0	2.0	2.0	2.0	2.0	2.0	2.0	2.0	2.0	2.0
Transfers SP	1.9	2.4	2.3	2.3	2.3	2.3	2.3	2.3	2.3	2.3	2.3	2.3	2.3
Interest	1.4	3.2	2.2	2.2	2.0	2.2	2.0	1.8	1.5	1.4	1.3	1.3	1.3
External	1.2	3.1	2.1	2.1	2.0	2.2	2.0	1.9	1.6	1.5	1.5	1.5	1.5
Domestic	0.2	0.1	0.2	0.1	0.0	0.0	-0.1	-0.1	-0.1	-0.1	-0.1	-0.2	-0.2
Exchange Rate Adjustments	4.6	0.5	0.0	0.0	0.0	0.0	0.0	0.0	0.0	0.0	0.0	0.0	0.0
Current Savings	-0.3	3.0	5.6	5.4	6.3	5.9	6.0	6.1	6.3	6.3	6.3	6.1	6.0
Capital Expenditures	7.4	6.8	5.0	4.9	5.1	5.1	5.1	5.2	5.2	5.3	5.3	5.3	5.3
Direct Investment	5.9	5.0	4.4	4.9	5.1	5.1	5.1	5.2	5.2	5.3	5.3	5.3	5.3
Indirect Investment	1.5	1.8	0.5	0.0	0.0	0.0	0.0	0.0	0.0	0.0	0.0	0.0	0.0
Deficit	7.6	3.9	-0.7	-0.5	-1.2	-0.8	-0.8	-0.8	-1.0	-1.0	-0.9	-0.8	-0.7
External Financing	2.3	1.5	1.5	2.0	0.4	0.1	0.3	0.5	0.1	0.2	0.2	0.2	0.2
Disbursements	2.3	1.5	1.5	2.6	12.5	2.4	2.2	2.3	2.8	2.8	2.9	3.0	3.1
Amortizations (program.)	0.0	0.0	0.0	2.9	2.5	2.3	1.9	1.8	2.7	2.7	2.7	2.7	2.9
Interest Arrears	0.0	0.0	0.0	0.5	-2.4	0.0	0.0	0.0	0.0	0.0	0.0	0.0	0.0
Amortization Arrears	0.0	0.0	0.0	1.8	-7.2	0.0	0.0	0.0	0.0	0.0	0.0	0.0	0.0
Domestic Financing	5.3	2.4	-2.1	-2.5	-1.5	-0.8	-1.1	-1.3	-1.2	-1.1	-1.1	-1.0	-0.8
Banks	-0.8	-2.4	-2.2	-1.6	-1.2	-0.6	-0.8	-1.0	-0.9	-0.8	-0.8	-0.8	-0.6
Other	6.1	4.9	0.1	-0.8	-0.4	-0.2	-0.3	-0.3	-0.3	-0.3	-0.3	-0.3	-0.2

Medium Case

Financial Sector (Bln. of Guaranies, stock)	1988	1989	1990	1991	1992	1993	1994	1995	1996	1997	1998	1999	2000
Total Assets	483.0	856.9	1231.3	1408.1	1692.2	1985.7	2292.8	2651.2	3057.2	3527.1	4075.3	4703.1	5434.0
Net International Reserves	287.7	536.7	891.9	1275.9	1334.4	1423.9	1524.9	1663.4	1784.1	1920.1	2074.8	2247.0	2436.3
Domestic Credit	368.2	407.0	440.1	232.9	477.2	699.8	925.7	1168.0	1478.6	1841.2	2267.8	2761.0	3345.5
Public	9.3	-103.4	-247.7	-370.0	-473.9	-538.7	-637.4	-772.2	-905.7	-1054.4	-1222.0	-1400.0	-1561.2
Private	358.9	510.3	687.7	602.9	951.1	1238.5	1563.1	1940.2	2384.2	2895.7	3489.8	4161.0	4906.7
Exchange Losses	100.6	13.8	0.0	0.0	0.0	0.0	0.0	0.0	0.0	0.0	0.0	0.0	0.0
Other Net Assets	-273.5	-100.6	-100.6	-100.6	-119.4	-138.1	-157.7	-180.2	-205.4	-234.3	-267.3	-304.9	-347.9
Total Liabilities	483.0	856.9	1146.3	1408.1	1692.20	1985.7	2292.8	2651.2	3057.2	3527.1	4075.3	4703.1	5434.0
M1	263.1	384.3	490.7	570.6	677.6	783.3	894.7	1022.0	1165.2	1328.9	1516.5	1729.6	1973.5
Currency	149.1	216.2	300.5	349.5	415.0	479.7	548.0	625.9	713.7	813.9	928.9	1059.4	1208.7
Demand Deposits	114.0	168.1	190.1	221.1	262.6	303.5	346.7	396.0	451.5	515.0	587.7	670.2	764.7
QM	185.6	403.6	586.6	768.6	945.6	1133.4	1329.2	1560.3	1823.0	2129.3	2489.8	2904.5	3391.5
Domestic Currency	167.6	225.2	290.9	413.6	509.0	609.1	719.3	848.6	998.3	1173.6	1379.4	1618.8	1899.2
Foreign Currency	18.0	178.4	295.7	355.0	436.6	524.3	609.8	711.6	824.7	955.6	1110.4	1285.7	1492.3
Foreign Liabilities LT	34.3	69.0	69.0	69.0	69.0	69.0	69.0	69.0	69.0	69.0	69.0	69.0	69.0
Memo items:													
Net Internatl. Reserves (Millions US$)	342.5	479.2	725.1	1025.1	1066.4	1122.9	1181.7	1258.2	1321.7	1390.2	1464.7	1544.1	1627.6
Foreign Liabilities L.P. (Millions US$)													
Unaccounted Funds (Millions US$)	-199.8	29.1	-674.3	-166.2	336.0	-49.3	-45.6	-32.5	-16.9	-8.6	2.7	16.9	29.3
Financial Sector (Billions 1982 Guaranies, stock)													
Total Assets	126.7	167.3	176.4	175.4	186.5	200.8	212.7	225.6	238.7	252.7	267.8	283.6	300.6
Net International Reserves	75.4	104.8	127.7	158.9	147.1	144.0	141.5	141.6	139.3	137.5	136.4	135.5	134.8
Domestic Credit	96.6	79.4	63.0	29.0	52.6	70.8	85.9	99.4	115.4	131.9	149.0	166.5	185.1
Public	2.4	-20.2	-35.5	-46.1	-52.2	-54.5	-59.1	-65.7	-70.7	-75.5	-80.3	-84.4	-86.4
Private	94.1	99.6	98.5	75.1	104.8	125.2	145.0	165.1	186.2	207.4	229.3	250.9	271.4
Exchange Losses	26.4	2.7	0.0	0.0	0.0	0.0	0.0	0.0	0.0	0.0	0.0	0.0	0.0
Other Net Assets	-71.7	-19.6	-14.4	-12.5	-13.2	-14.0	-14.6	-15.3	-16.0	-16.8	-17.6	-18.4	-19.2
Total Liabilites	126.7	167.3	164.2	175.4	186.5	200.8	212.7	225.6	238.7	252.7	267.8	283.6	300.6
M1	69.0	75.0	70.3	71.1	74.7	79.2	83.0	87.0	91.0	95.2	99.7	104.3	109.2
Currency	39.1	42.2	43.0	43.5	45.7	48.5	50.8	53.3	55.7	58.3	61.0	63.9	66.9
Demand Deposits	29.9	32.8	27.2	27.5	28.9	30.7	32.2	33.7	35.3	36.9	38.6	40.4	42.3
QM	48.7	78.8	84.0	95.7	104.2	114.6	123.3	132.8	142.3	152.5	163.6	175.1	187.6
Domestic Currency	44.0	44.0	41.7	51.5	56.1	61.6	66.7	72.2	78.0	84.1	90.7	97.6	105.1
Foreign Currency	4.7	34.8	42.4	44.2	48.1	53.0	56.6	60.6	64.4	68.5	73.0	77.5	82.5
Foreign Liabilities LT	9.0	13.5	9.9	8.6	7.6	7.0	6.4	5.9	5.4	4.9	4.5	4.2	3.8

Medium Case

- 115 -

Debt (Pipeline + New) (Mln. US$)	1988	1989	1990	1991	1992	1993	1994	1995	1996	1997	1998	1999	2000
Total Commitments	211.8	0.0	0.0	224.8	822.5	324.5	296.6	321.2	418.6	462.0	512.9	572.4	641.6
Refinancing	0.0	0.0	0.0	0.0	569.5	0.0	0.0	0.0	0.0	0.0	0.0	0.0	0.0
Other Multilateral and IDB	79.9	0.0	0.0	50.0	52.9	55.5	59.3	64.5	70.1	76.2	82.8	90.0	97.9
World Bank	0.0	0.0	0.0	5.0	30.0	100.0	53.4	58.0	63.1	68.6	74.6	81.0	88.1
Brazil (CACEX)	0.0	0.0	0.0	0.0	0.0	0.0	0.0	0.0	0.0	0.0	0.0	0.0	0.0
Bilateral	93.6	0.0	0.0	70.0	80.0	90.0	96.1	104.5	113.6	123.5	134.2	145.9	158.6
Suppliers + Export Credits	0.0	0.0	0.0	0.0	0.0	0.0	0.0	0.0	0.0	0.0	0.0	0.0	0.0
Financial Institutions	38.3	0.0	0.0	0.0	0.0	0.0	0.0	0.0	0.0	0.0	0.0	0.0	0.0
Private Non Guaranteed	0.0	0.0	0.0	0.0	0.0	0.0	0.0	0.0	0.0	0.0	0.0	0.0	0.0
IMF	0.0	0.0	0.0	0.0	0.0	0.0	0.0	0.0	0.0	0.0	0.0	0.0	0.0
Short Term	0.0	0.0	0.0	70.0	70.6	70.6	71.8	74.5	77.2	80.1	83.1	86.1	89.3
Unidentified Sources	0.0	0.0	0.0	29.8	19.5	8.4	15.9	19.7	94.6	113.6	138.3	169.3	207.8
Total Disbursements	141.8	628.8	80.0	240.7	805.5	246.7	241.8	280.6	373.9	407.8	454.9	512.1	576.6
Refinancing	0.0	0.0	0.0	0.0	569.5	0.0	0.0	0.0	0.0	0.0	0.0	0.0	0.0
Other Multilateral and IDB	31.4	18.0	15.0	24.6	33.6	44.2	42.0	54.4	59.3	63.6	65.8	71.6	77.5
World Bank	18.0	16.1	16.0	12.3	10.6	34.6	40.2	23.8	33.7	43.5	52.4	59.8	65.8
Brazil (CACEX)	0.0	426.6	0.0	0.0	0.0	0.0	0.0	0.0	0.0	0.0	0.0	0.0	0.0
Bilateral	40.4	125.9	22.0	70.7	81.4	75.2	71.9	99.3	109.1	107.0	115.3	125.3	136.2
Suppliers + Export Credits	14.8	41.2	25.0	25.6	15.2	10.1	0.0	2.6	0.0	0.0	0.0	0.0	0.0
Financial Institutions	37.2	0.9	2.0	7.6	5.2	3.6	0.0	6.3	0.0	0.0	0.0	0.0	0.0
Private Non Guaranteed	0.0	0.0	0.0	0.0	0.0	0.0	0.0	0.0	0.0	0.0	0.0	0.0	0.0
IMF	0.0	0.0	0.0	0.0	0.0	0.0	0.0	0.0	0.0	0.0	0.0	0.0	0.0
Short Term	-63.0	44.0	76.0	70.0	70.6	70.6	71.8	74.5	77.2	80.1	83.1	86.1	89.3
Unidentified Sources	0.0	0.0	0.0	29.8	19.5	8.4	15.9	19.7	94.6	113.6	138.3	169.3	207.8
Amortization Arrears	44.4	13.9	122.2	100.0	-425.0	0.0	0.0	0.0	0.0	0.0	0.0	0.0	0.0
Total Amortizations	180.8	449.8	533.9	203.4	185.7	179.2	157.3	165.8	268.7	288.1	318.2	353.4	411.1
Refinancing	0.0	0.0	0.0	0.0	0.0	0.0	0.0	0.0	81.4	81.4	81.4	81.4	81.4
Other Multilateral and IDB	13.0	22.4	21.0	27.1	29.3	30.8	31.8	40.4	49.3	60.6	67.2	77.9	89.6
World Bank	42.8	37.7	41.0	38.5	38.7	38.2	37.2	32.9	30.8	24.7	24.4	28.1	39.4
Brazil (CACEX)	0.0	0.0	426.6	0.0	0.0	0.0	0.0	0.0	0.0	6.4	6.4	6.4	6.4
Bilateral	41.1	59.5	28.0	44.1	44.9	45.7	42.2	59.4	75.5	91.6	113.2	130.6	149.6
Suppliers + Export Credits	40.3	5.7	16.0	42.5	38.1	33.0	29.6	17.9	15.6	8.5	8.5	8.5	8.5
Financial Institutions	43.6	323.1	0.0	49.6	32.9	29.3	14.8	8.3	6.0	3.9	3.9	3.9	3.9
Private Non Guaranteed	0.0	1.3	1.3	1.6	1.8	2.1	1.7	1.8	1.9	1.4	1.0	1.0	1.0
IMF	0.0	0.0	0.0	0.0	0.0	0.0	0.0	0.0	0.0	0.0	0.0	0.0	0.0
Unidentified Sources	0.0	0.0	0.0	0.0	0.0	0.0	0.0	5.0	8.2	9.6	12.3	15.6	31.3

Medium Case

Debt (Pipeline + New) (Mln. US$) cont.	1988	1989	1990	1991	1992	1993	1994	1995	1996	1997	1998	1999	2000
Interest Arrears	15.7	-14.7	36.0	30.0	-144.5	0.0	0.0	0.0	0.0	0.0	0.0	0.0	0.0
Total Interest	114.6	91.4	90.0	108.2	107.1	126.8	123.5	124.5	116.2	117.7	124.6	137.9	152.4
Refinancing	0.0	0.0	0.0	0.0	22.8	43.3	40.4	39.9	33.8	28.6	23.4	18.2	13.0
Other Multilateral and IDB	20.6	20.5	22.0	21.3	21.0	21.3	21.9	22.5	22.3	22.4	22.4	24.0	25.3
World Bank	30.2	25.6	25.0	22.2	20.1	18.8	18.2	17.3	15.9	15.7	16.3	19.6	22.9
Brazil (CACEX)	0.1	14.8	0.0	0.0	0.0	0.0	0.0	0.0	0.0	-0.2	-0.3	-0.5	-0.6
Bilateral	17.7	9.2	10.0	16.3	17.1	18.3	19.3	20.5	19.6	19.5	22.4	24.9	27.0
Suppliers + Export Credits	25.0	7.0	17.0	12.6	11.2	8.7	6.2	3.9	2.3	1.4	0.7	0.7	0.7
Financial Institutions	19.4	2.5	1.0	10.9	7.2	4.5	2.7	1.6	1.0	0.5	0.2	0.2	0.2
Private Non Guaranteed	0.7	1.1	0.0	1.0	1.0	0.9	0.9	0.9	0.9	0.9	0.8	0.8	0.8
Interest on Arrears	0.0	9.8	14.0	22.8	0.0	0.0	0.0	0.0	0.0	0.0	0.0	0.0	0.0
IMF	0.0	0.0	0.0	0.0	0.0	0.0	0.0	0.0	0.0	0.0	0.0	0.0	0.0
Short Term	1.0	1.0	1.0	1.0	4.3	7.3	9.9	12.7	14.8	17.7	20.8	24.0	27.3
Unidentified Sources	0.0	0.0	0.0	0.0	2.4	3.8	4.1	5.2	5.7	11.2	17.8	25.9	35.7
Other Interest													
Accumulated Arrears	282.1	281.3	439.5	569.5	0.0	0.0	0.0	0.0	0.0	0.0	0.0	0.0	0.0
Total DOD	2354.9	2388.7	2127.5	2294.9	2345.2	2412.7	2497.2	2612.1	2717.4	2837.0	2973.7	3132.4	3297.9
Refinancing	0.0	0.0	0.0	0.0	569.5	569.5	569.5	569.5	488.1	406.8	325.4	244.1	162.7
Other Multilateral and IDB	419.4	409.7	415.0	412.5	416.9	430.3	440.5	454.4	464.5	467.5	466.1	459.7	457.6
World Bank	358.1	324.3	320.0	293.8	265.6	262.1	265.1	256.0	258.9	277.8	305.8	337.6	364.0
Brazil (CACEX)	0.0	426.6	0.0	0.0	0.0	0.0	0.0	0.0	0.0	-6.4	-12.8	-19.2	-25.6
Bilateral	478.2	450.4	470.0	496.7	533.1	562.5	592.3	632.1	665.7	681.0	683.2	677.8	664.5
Suppliers + Export credits	210.3	219.3	248.0	231.1	208.2	185.3	155.7	140.4	124.7	116.2	107.7	99.2	90.7
Financial Institutions	629.8	269.9	283.0	241.0	213.3	187.6	172.8	170.8	164.8	160.8	156.9	153.0	149.0
Private Non Guaranteed	28.0	28.0	19.0	17.4	15.6	13.5	11.8	10.0	8.1	6.7	5.7	4.6	3.6
IMF	0.0	0.0	0.0	0.0	0.0	0.0	0.0	0.0	0.0	0.0	0.0	0.0	0.0
Unidentified Sources	0.0	0.0	0.0	29.8	49.4	57.8	73.6	88.4	174.8	278.8	404.8	558.5	735.0
Short Term	231.2	260.5	372.5	472.5	398.6	469.1	541.0	615.4	692.7	772.8	855.8	942.0	1031.3
Interest Arrears	93.2	78.5	114.5	144.5	0.0	0.0	0.0	0.0	0.0	0.0	0.0	0.0	0.0
Other	138.0	182.0	258.0	328.0	398.6	469.1	541.0	615.4	692.7	772.8	855.8	942.0	1031.3
Adjustment in Debt Stock	-2.9	-3.7	-3.7	-3.7	-3.7	-3.7	-3.7	-3.7	-3.7	-3.7	-3.7	-3.7	-3.7
Net Disbursements:													
Other Multilateral and IDB	18.4	-4.4	-6.0	-2.5	4.3	13.4	10.2	14.0	10.1	3.0	-1.4	-6.4	-12.1
World Bank	-24.9	-21.6	-25.0	-26.2	-28.1	-3.6	3.0	-9.1	2.9	18.8	28.1	31.7	26.4
Brazil (CACEX)	0.0	426.6	-426.6	0.0	0.0	0.0	0.0	0.0	0.0	-6.4	-6.4	-6.4	-6.4
Bilateral	-0.7	66.4	-6.0	26.7	36.4	29.5	29.7	39.8	33.6	15.3	2.1	-5.3	-13.4
Memorandum:													
Total Debt	2352.0	2385.0	2123.8	2291.2	2341.5	2409.0	2493.5	2608.4	2713.7	2833.3	2970.0	3128.7	3294.2
Debt Service/Exports (%)	37.3	41.2	43.3	23.6	14.4	13.6	11.6	10.9	13.3	13.0	13.0	13.3	14.0
Debt Service/GDP (%)	11.0	15.4	16.3	9.8	4.9	4.9	4.2	4.0	4.9	4.7	4.7	4.8	5.0
Interest/Exports (%)	11.2	5.0	7.0	7.4	5.3	5.7	5.1	4.7	4.0	3.8	3.7	3.7	3.8
Interest/GDP (%)	3.3	1.9	2.4	2.4	1.8	2.0	1.9	1.7	1.5	1.4	1.3	1.3	1.4
Debt/Exports (%)	227.4	173.5	142.3	153.0	115.0	107.5	103.3	98.5	94.3	90.6	87.3	84.7	82.0
Debt/GDP (%)	66.7	64.9	48.8	50.6	39.5	38.7	37.4	36.0	34.3	32.8	31.6	30.4	29.4
Medium Case													

Paraguay Projections Model

8:50 AM
26-Sep-91

	1988	1989	1990	1991	1992	1993	1994	1995	1996	1997	1998	1999	2000
Population (Thousands)	4039.2	4157.3	4278.9	4404.0	4532.8	4665.4	4801.8	4942.2	5086.8	5235.5	5388.6	5546.2	5708.4
Parallel Exchange Rate (G/$, base year)	136.0	136.0	136.0	136.0	136.0	136.0	136.0	136.0	136.0	136.0	136.0	136.0	136.0

Investment Financing (Bln. 1982 Guaranies)

	1988	1989	1990	1991	1992	1993	1994	1995	1996	1997	1998	1999	2000
Total Savings	201.4	204.1	204.2	212.9	232.7	252.3	273.0	294.2	317.0	341.8	368.9	398.3	430.3
Private Savings	143.4	201.2	142.7	143.6	154.2	177.1	191.4	204.4	219.9	237.5	257.3	279.0	305.0
Financial Savings	-5.8	36.1	0.5	12.5	22.4	25.2	24.2	26.7	28.6	31.9	35.8	39.2	44.1
Non-financial Savings	149.2	165.1	142.2	131.1	131.9	151.9	167.2	177.7	191.3	205.5	221.5	239.8	260.9
Public Savings	-2.2	26.7	52.2	50.9	65.1	65.9	71.2	77.6	85.3	91.5	97.4	102.4	107.7
Foreign Savings	44.6	-47.6	-18.0	7.8	6.3	5.6	6.5	7.9	8.4	9.2	10.4	12.7	13.2
Inflation Tax	15.6	23.7	27.2	10.5	7.1	3.6	3.9	4.2	3.4	3.6	3.9	4.1	4.4
Total Investment	201.4	204.1	204.2	212.9	232.7	252.3	273.0	294.2	317.0	341.8	368.9	398.3	430.3
Private Investment	137.1	142.6	158.0	166.6	180.2	194.9	210.7	226.7	244.0	262.8	283.2	305.4	329.5
Public Investment	64.3	61.5	46.2	46.4	52.6	57.4	62.2	67.5	73.0	79.1	85.7	92.9	100.8

Investment Financing (% of GDP)

	1988	1989	1990	1991	1992	1993	1994	1995	1996	1997	1998	1999	2000
Total Savings	23.1	22.1	21.9	22.7	23.2	23.5	23.9	24.2	24.5	24.8	25.1	25.5	25.8
Private Savings	16.5	21.8	15.3	15.3	15.4	16.5	16.7	16.8	17.0	17.2	17.5	17.9	18.3
Financial Savings	-0.7	3.9	0.1	1.3	2.2	2.4	2.1	2.2	2.2	2.3	2.4	2.5	2.6
Non-financial Savings	17.1	17.9	15.2	14.0	13.2	14.1	14.6	14.6	14.8	14.9	15.1	15.3	15.6
Public Savings	-0.3	2.9	5.6	5.4	6.5	6.1	6.2	6.4	6.6	6.6	6.6	6.6	6.5
Foreign Savings	5.1	-5.2	-1.9	0.8	0.6	0.5	0.6	0.7	0.6	0.7	0.7	0.8	0.8
Inflation Tax	1.8	2.6	2.9	1.1	0.7	0.3	0.3	0.3	0.3	0.3	0.3	0.3	0.3
Total Investment	23.1	22.1	21.9	22.7	23.2	23.5	23.9	24.2	24.5	24.8	25.1	25.5	25.8
Private Investment	15.8	15.4	16.9	17.8	18.0	18.1	18.4	18.6	18.9	19.1	19.3	19.5	19.8
Public Investment	7.4	6.7	4.9	4.9	5.2	5.3	5.4	5.5	5.6	5.7	5.8	5.9	6.0

Memo items:

	1988	1989	1990	1991	1992	1993	1994	1995	1996	1997	1998	1999	2000
Average Propensity to Save	17.8	25.4	18.0	17.5	17.7	18.9	19.2	19.3	19.6	19.9	20.3	20.7	21.2
Marginal Propensity to Save	90.9	-397.5	-3015.8	3.1	21.9	35.5	24.2	21.0	23.8	25.0	26.0	26.6	29.2

High Case

- 118 -

National Accounts (Bln. 1982 Guaranies)	1988	1989	1990	1991	1992	1993	1994	1995	1996	1997	1998	1999	2000
Total Consumption	728.1	703.9	721.5	737.0	781.3	831.2	881.4	936.1	991.8	1051.4	1115.7	1184.1	1256.9
Private Consumption	686.1	647.8	660.9	675.1	713.1	755.9	800.0	848.3	897.3	949.5	1005.7	1065.3	1128.4
Public Consumption	42.0	56.2	60.5	61.9	68.2	75.4	81.3	87.7	94.5	101.9	110.0	118.8	128.4
Total Fixed Investment	167.0	184.8	203.4	212.1	231.9	251.3	272.0	293.1	315.8	340.6	367.5	396.8	428.7
Private Investment	113.6	129.1	157.4	165.9	179.5	194.1	210.0	225.8	243.1	261.8	282.1	304.2	328.3
Public Investment	53.3	55.7	46.0	46.2	52.4	57.2	62.0	67.2	72.7	78.8	85.4	92.6	100.4
Exports	157.1	196.5	229.8	234.6	256.2	281.1	301.7	323.4	344.7	367.7	392.6	418.6	446.8
Imports	201.9	185.7	227.3	228.6	253.6	283.5	306.0	330.0	350.6	372.6	396.8	421.0	446.7
GDP	850.2	899.5	927.3	955.1	1015.7	1080.2	1149.0	1222.5	1301.7	1387.1	1479.1	1578.5	1685.7
Terms-of-trade Adjustment	20.2	23.7	6.1	-17.4	-13.4	-6.1	-5.7	-5.5	-7.9	-10.3	-12.2	-15.3	-18.0
GDY	870.4	923.2	933.4	937.7	1002.3	1074.1	1143.3	1217.0	1293.8	1376.8	1467.0	1563.2	1667.7
Transfers	5.4	3.4	4.3	4.2	4.2	4.3	4.2	4.2	4.2	4.2	4.2	4.1	4.1
Net Factor Income	-11.7	-6.8	0.5	1.7	2.3	0.4	1.3	2.3	3.8	4.6	5.0	4.7	4.5
GNY	864.1	919.8	938.2	943.6	1008.8	1078.7	1148.8	1223.6	1301.8	1385.6	1476.1	1572.0	1676.3
GNP	843.9	896.2	932.1	961.0	1022.2	1084.8	1154.6	1229.1	1309.7	1395.9	1488.2	1587.3	1694.3
P GDY Deflator (1982=1) a/	3.81	5.12	6.98	8.03	8.83	9.26	9.70	10.17	10.54	10.93	11.34	11.76	12.19
Inflation	23.10%	34.35%	36.28%	15.00%	10.00%	4.80%	4.80%	4.80%	3.70%	3.70%	3.70%	3.70%	3.70%
P Investment Deflator (1982=1)	4.60	5.66	7.01	8.06	8.87	9.29	9.74	10.20	10.58	10.97	11.38	11.80	12.24
P Inv. Deflator / P GDY Deflator	1.21	1.10	1.00	1.00	1.00	1.00	1.00	1.00	1.00	1.00	1.00	1.00	1.00
P Consumption Deflator	3.69	4.68	6.89	8.07	8.88	9.29	9.74	10.21	10.59	10.99	11.40	11.82	12.25
ICOR	3.17	3.39	6.64	7.31	3.50	3.60	3.65	3.70	3.70	3.70	3.70	3.70	3.70
MUV	1.34	1.35	1.43	1.57	1.58	1.58	1.61	1.67	1.73	1.80	1.86	1.93	2.00
GDP Growth	6.4	5.8	3.1	3.0	6.3	6.3	6.4	6.4	6.5	6.6	6.6	6.7	6.8
Private Consumption Growth	3.7	-5.6	2.0	2.1	5.6	6.0	5.8	6.0	5.8	5.8	5.9	5.9	5.9
National Accounts (Bln. Current Guaranies)	1988	1989	1990	1991	1992	1993	1994	1995	1996	1997	1998	1999	2000
Total Consumption	2686.4	3292.9	4968.5	5950.3	6935.3	7719.9	8582.4	9558.6	10507.3	11553.4	12714.4	13994.1	15398.4
Private Consumption	2531.3	3030.1	4551.5	5450.4	6329.7	7020.0	7790.5	8662.9	9506.1	10433.5	11460.4	12590.1	13824.7
Public Consumption	155.1	262.7	417.0	500.0	605.5	699.9	791.9	895.8	1001.2	1119.9	1254.0	1404.0	1573.7
Total Fixed Investment	768.2	1045.6	1425.4	1709.6	2055.6	2334.9	2648.1	2990.3	3341.9	3737.1	4182.1	4682.1	5245.6
Private Investment	522.8	730.5	1103.0	1337.3	1591.4	1803.7	2044.3	2304.5	2572.1	2872.6	3210.3	3589.9	4016.9
Public Investment	245.4	315.1	322.4	372.2	464.2	531.2	603.7	685.9	769.8	864.5	971.8	1092.2	1228.7
Exports	974.1	1723.2	2219.5	2490.1	3083.1	3660.9	4129.0	4647.5	5105.6	5618.7	6202.7	6818.2	7518.1
Imports	1109.5	1453.3	2139.0	2620.9	3221.6	3774.2	4269.1	4824.2	5315.3	5857.5	6468.9	7117.7	7831.0
GDP	3319.1	4608.4	6474.4	7529.1	8852.4	9941.6	11090.4	12372.3	13639.5	15051.7	16630.3	18376.7	20331.1
External Transfers	29.6	26.8	40.5	47.9	53.6	56.6	59.1	61.7	63.6	65.5	67.7	69.7	71.9
Net Factor Income	-64.2	-52.9	4.8	19.9	28.9	4.7	18.3	34.2	57.7	72.6	80.8	80.3	79.5
GNP	3254.9	4555.5	6479.3	7549.0	8881.3	9946.3	11108.7	12406.5	13697.1	15124.3	16711.1	18457.0	20410.6
GDP US$	3951.3	4114.6	5263.8	5661.0	6002.2	6374.7	6939.4	7685.1	8531.0	9470.4	10507.2	11688.9	12996.0
GNP US$	3874.9	4067.4	5267.7	5675.9	6021.8	6377.7	6950.8	7706.3	8567.1	9516.1	10558.2	11739.9	13046.8
Per capita GDP (US$)	978.3	989.7	1230.2	1285.4	1324.2	1366.4	1445.2	1555.0	1677.1	1808.9	1949.9	2107.5	2276.6
Per capita GNP (US$)	959.3	978.4	1231.1	1288.8	1328.5	1367.0	1447.5	1559.3	1684.2	1817.6	1959.3	2116.7	2285.5

a/ Current GDP/Constant GDY; 1+Inflation

High Case

- 119 -

The Balance of Payments (Millions US$)	1988	1989	1990	1991	1992	1993	1994	1995	1996	1997	1998	1999	2000	
Exports	1159.6	1538.6	1804.5	1872.3	2090.4	2347.5	2583.6	2886.8	3193.4	3535.2	3918.9	4336.9	4805.7	94
Goods	871.0	1166.5	1392.3	1449.2	1617.6	1818.0	2001.4	2236.5	2474.4	2739.5	3037.2	3361.4	3725.2	95
Non-factor Services	288.6	372.1	412.2	423.1	472.8	529.5	582.2	650.3	719.0	795.7	881.7	975.4	1080.5	96
														97
Imports	1320.8	1297.6	1739.0	1970.6	2184.3	2420.1	2671.2	2996.5	3324.5	3685.5	4087.1	4527.4	5005.7	98
Goods	1030.1	1001.3	1353.6	1583.7	1747.5	1924.5	2118.7	2370.6	2627.0	2909.1	3222.1	3567.9	3942.7	99
Non-factor Services	290.7	296.3	385.4	386.9	436.9	495.6	552.6	625.9	697.5	776.4	865.0	959.4	1063.0	100
														101
Resource Balance	-161.2	241.0	65.5	-98.3	-93.9	-72.6	-87.6	-109.8	-131.2	-150.3	-168.1	-190.5	-200.0	102
														103
Factor Services (net)	-76.4	-47.2	3.9	14.9	19.6	3.0	11.4	21.2	36.1	45.7	51.0	51.1	50.8	104
Interest Payments	-130.3	-130.9	-90.0	-108.2	-107.1	-127.4	-123.9	-124.0	-114.7	-114.8	-120.0	-131.3	-143.7	105
Others	53.9	83.7	93.9	123.1	126.7	130.5	135.3	145.2	150.7	160.5	171.1	182.4	194.5	106
														107
Transfers	35.2	23.9	32.9	36.0	36.3	36.3	37.0	38.3	39.8	41.2	42.7	44.3	46.0	108
														109
Current Account	-202.4	217.7	102.3	-47.4	-38.0	-33.3	-39.2	-50.2	-55.3	-63.4	-74.4	-95.1	-103.2	110
														111
Capital Account	-1.5	-80.2	-14.6	217.4	660.9	102.0	112.5	145.1	137.3	153.6	174.8	205.1	222.8	112
														113
Direct Foreign Investment	0.0	0.0	73.1	30.0	33.3	36.6	41.0	46.7	53.3	60.8	69.4	79.1	90.3	114
														115
Net MLT Flows	-83.4	165.1	-576.1	-62.5	529.7	-11.4	-3.2	20.6	-67.9	-75.0	-83.5	-92.8	-124.3	116
Disbursements	141.8	628.8	80.0	140.9	715.4	167.8	154.1	186.4	202.1	214.1	233.5	256.7	279.5	117
Amortizations	225.2	463.7	656.1	203.4	185.7	179.2	157.3	165.8	270.0	289.1	317.0	349.5	403.8	118
														119
Other Capital Flows	-99.9	-4.6	352.0	70.0	70.6	70.6	71.8	74.5	77.2	80.1	83.1	86.1	89.3	120
														121
Other	181.8	-240.7	136.4	150.0	0.0	0.0	0.0	0.0	0.0	0.0	0.0	0.0	0.0	122
														123
Financing Gap	0.0	0.0	0.0	29.8	27.4	6.3	2.8	3.2	74.6	87.7	105.8	132.7	167.5	124
														125
Overall BOP Surplus	-203.9	137.5	87.7	170.0	622.9	68.8	73.2	94.9	82.0	90.2	100.4	110.1	119.6	126
Change in Net International Reserves	-143.8	136.7	245.9	300.0	53.4	68.8	73.2	94.9	82.0	90.2	100.4	110.1	119.6	127
Arrears	60.1	-0.8	158.2	130.0	-569.5	0.0	0.0	0.0	0.0	0.0	0.0	0.0	0.0	128
Amortizations	44.4	13.9	122.2	100.0	-425.0	0.0	0.0	0.0	0.0	0.0	0.0	0.0	0.0	129
Interest	15.7	-14.7	36.0	30.0	-144.5	0.0	0.0	0.0	0.0	0.0	0.0	0.0	0.0	130
														131
Memo items:														132
Nominal Exchange Rate (Average)	840.0	1120.0	1230.0	1330.0	1474.9	1559.5	1598.2	1609.9	1598.8	1559.3	1582.8	1572.2	1564.4	133
Real Exchange Rate	220.3	218.6	176.2	165.6	167.0	168.5	164.8	158.4	151.7	145.4	139.6	133.7	128.3	134
Implicit Exchange Rate on Imports		218.6	176.2	165.6	167.0	168.5	164.8	158.4	151.7	145.4	139.6	133.7	128.3	135
QR Policy Variable	1	1	1	1	1	1	1	1	1	1	1	1	1	136
Real Eff. Exch. Rate (1982=1)	1.4411	1.5276	1.3477	1.4281	1.4381	1.4381	1.4381	1.4381	1.4381	1.4381	1.4381	1.4381	1.4381	137
Amortization (DRS)	236.7													138
BOP Interest - DRS Interest	-77.5	-39	-114.5	-144.5	0	0	0	0	0	0	0	0	0	139
Terms of Trade (1982=100)	116.1	113.6	97.9	92.5	94.8	98.1	98.2	98.3	97.6	96.9	96.5	95.8	95.2	140
Gap plus Net MLT	-83.4	165.1	-576.1	-32.6	557.1	-5.1	-0.3	23.9	6.8	12.7	22.3	39.9	43.2	141
Net Binationals														142
														143
Stock of Int'l. Reserves	291.2	427.9	673.8	973.8	1027.2	1096.0	1169.2	1264.1	1346.1	1436.4	1536.8	1646.8	1766.4	144

High Case

- 120 -

Public Sector (% of GDP)	1988	1989	1990	1991	1992	1993	1994	1995	1996	1997	1998	1999	2000	
Current Revenues	14.3	17.3	18.9	19.0	20.0	20.2	20.2	20.2	20.2	20.2	20.2	20.2	20.2	308
Taxes	7.0	8.1	9.2	9.3	10.3	10.5	10.5	10.5	10.5	10.5	10.5	10.5	10.5	309
Imports	1.1	1.6	2.3	2.6	3.0	3.1	3.2	3.2	3.2	3.2	3.2	3.2	3.2	310
Exports	0.0	0.8	0.5	0.0	0.0	0.0	0.0	0.0	0.0	0.0	0.0	0.0	0.0	311
Direct	2.2	1.8	1.5	1.7	1.8	1.8	1.8	1.8	1.8	1.8	1.8	1.8	1.8	312
Indirect	3.7	3.9	5.0	5.0	5.5	5.5	5.5	5.5	5.5	5.5	5.5	5.5	5.5	313
Non-tax Revenues	3.8	6.1	5.4	5.4	5.4	5.4	5.4	5.4	5.4	5.4	5.4	5.4	5.4	314
V.A. on Public Enterprises	3.4	3.1	4.3	4.3	4.3	4.3	4.3	4.3	4.3	4.3	4.3	4.3	4.3	315
														316
														317
														318
Current Expenditures	10.0	13.8	13.3	13.5	13.5	14.0	14.0	13.9	13.6	13.6	13.6	13.7	13.7	319
Government Salaries	2.9	3.8	4.4	4.6	4.8	5.0	5.1	5.2	5.3	5.4	5.5	5.6	5.7	320
Public Enterprise Salaries	1.5	1.5	1.5	1.5	1.5	1.6	1.7	1.7	1.7	1.7	1.7	1.7	1.7	321
Social Security & PE Tax	0.6	0.9	0.9	0.9	0.9	0.9	0.9	0.9	0.9	0.9	0.9	0.9	0.9	322
Other Government	1.7	1.9	2.0	2.0	2.0	2.0	2.0	2.0	2.0	2.0	2.0	2.0	2.0	323
Transfers SP	1.9	2.4	2.3	2.3	2.3	2.3	2.3	2.3	2.3	2.3	2.3	2.3	2.3	324
Interest	1.4	3.2	2.2	2.2	2.0	2.1	1.9	1.7	1.4	1.2	1.1	1.1	1.1	325
External	1.2	3.1	2.1	2.1	2.0	2.2	2.0	1.8	1.5	1.3	1.3	1.2	1.2	326
Domestic	0.2	0.1	0.2	0.1	0.0	-0.1	-0.1	-0.1	-0.1	-0.1	-0.1	-0.1	-0.2	327
														328
Exchange Rate Adjustments	4.6	0.5	0.0	0.0	0.0	0.0	0.0	0.0	0.0	0.0	0.0	0.0	0.0	329
														330
Current Savings	-0.3	3.0	5.6	5.4	6.5	6.1	6.2	6.4	6.6	6.6	6.6	6.6	6.5	331
														332
Capital Expenditures	7.4	6.8	5.0	4.9	5.2	5.3	5.4	5.5	5.6	5.7	5.8	5.9	6.0	333
Direct Investment	5.9	5.0	4.4	4.9	5.2	5.3	5.4	5.5	5.6	5.7	5.8	5.9	6.0	334
Indirect Investment	1.5	1.8	0.5	0.0	0.0	0.0	0.0	0.0	0.0	0.0	0.0	0.0	0.0	335
														336
Deficit	7.6	3.9	-0.7	-0.5	-1.2	-0.8	-0.8	-0.8	-1.0	-0.9	-0.8	-0.6	-0.4	337
External Financing	2.3	1.5	1.5	2.0	0.5	0.1	0.2	0.3	0.0	0.0	0.0	0.0	0.0	338
Disbursements	2.3	1.5	1.5	2.6	12.5	2.3	2.0	2.1	2.5	2.4	2.4	2.4	2.5	339
Amortizations (program.)	0.0	0.0	0.0	2.9	2.5	2.2	1.8	1.7	2.5	2.4	2.4	2.4	2.5	340
Interest Arrears	0.0	0.0	0.0	0.5	-2.4	0.0	0.0	0.0	0.0	0.0	0.0	0.0	0.0	341
Amortization Arrears	0.0	0.0	0.0	1.8	-7.1	0.0	0.0	0.0	0.0	0.0	0.0	0.0	0.0	342
Domestic Financing	5.3	2.4	-2.1	-2.5	-1.8	-0.8	-0.9	-1.2	-0.9	-0.9	-0.8	-0.7	-0.4	343
Banks	-0.8	-2.4	-2.2	-1.6	-1.3	-0.6	-0.7	-0.9	-0.7	-0.7	-0.6	-0.5	-0.3	344
Other	6.1	4.9	0.1	-0.8	-0.4	-0.2	-0.2	-0.3	-0.2	-0.2	-0.2	-0.2	-0.1	345
High Case														346

Financial Sector (Bln. of Guaranies, stock)	1988	1989	1990	1991	1992	1993	1994	1995	1996	1997	1998	1999	2000
Total Assets	483.0	856.9	1231.3	1408.1	1739.6	2053.4	2383.2	2765.9	3167.5	3631.0	4168.8	4781.4	5493.0
Net International Reserves	287.7	536.7	891.9	1275.5	1350.8	1455.1	1570.8	1723.0	1854.5	1998.4	2157.6	2331.2	2518.8
Domestic Credit	368.2	407.0	440.1	232.9	507.0	731.1	960.6	1208.2	1495.1	1833.7	2233.3	2695.7	3245.9
Public	9.3	-103.4	-247.7	-370.0	-486.4	-549.2	-628.2	-736.7	-829.7	-929.4	-1028.0	-1118.5	-1180.0
Private	358.9	510.3	687.7	602.9	993.5	1280.3	1588.8	1945.0	2324.8	2763.0	3261.3	3814.2	4425.9
Exchange Losses	100.6	13.8	0.0	0.0	0.0	0.0	0.0	0.0	0.0	0.0	0.0	0.0	0.0
Other Net Assets	-273.5	-100.6	-100.6	-100.6	-118.3	-132.8	-148.2	-165.3	-182.2	-201.1	-222.2	-245.5	-271.6
Total Liabilites	483.0	856.9	1146.3	1408.1	1739.55	2053.4	2383.2	2765.9	3167.5	3631.0	4168.8	4781.4	5493.0
M1	263.1	384.3	490.7	570.6	670.9	753.4	840.5	937.6	1033.6	1140.7	1260.3	1392.6	1540.8
Currency	149.1	216.2	300.5	349.5	410.9	461.5	514.8	574.3	633.1	698.7	771.9	853.0	943.7
Demand Deposits	114.0	168.1	190.1	221.1	260.0	292.0	325.7	363.3	400.5	442.0	488.4	539.7	597.1
QM	185.6	403.6	586.6	768.6	999.7	1231.0	1473.8	1759.6	2064.8	2421.3	2839.5	3319.8	3883.3
Domestic Currency	167.6	225.2	290.9	413.6	548.2	685.3	831.0	1001.3	1185.7	1398.8	1645.2	1928.3	2255.3
Foreign Currency	18.0	178.4	295.7	355.0	451.5	545.7	642.7	758.0	879.1	1022.6	1194.3	1391.5	1627.9
Foreign Liabilities LT	34.3	69.0	69.0	69.0	69.0	69.0	69.0	69.0	69.0	69.0	69.0	69.0	69.0
Memo Items:													
Net Internat. Reserves (Millions US$)	342.5	479.2	725.1	1025.1	1078.5	1147.3	1220.5	1315.4	1397.4	1487.7	1588.1	1698.1	1817.7
Foreign Liabilities L.P. (Millions US$)													
Unaccounted Funds (Millions US$)	-199.8	29.1	-674.3	-166.2	335.6	-52.8	-54.9	-44.8	-32.8	-28.6	-21.1	-9.7	0.4

Financial Sector (Billions 1982 Guaranies, stock)													
Total Assets	126.7	167.3	176.4	175.4	197.0	221.8	245.7	272.1	300.5	332.1	367.7	406.7	450.6
Net International Reserves	75.4	104.8	127.7	158.9	152.9	157.2	161.9	169.5	175.9	182.8	190.3	198.3	206.6
Domestic Credit	96.6	79.4	63.0	29.0	57.4	79.0	99.0	118.8	141.8	167.7	197.0	229.3	266.3
Public	2.4	-20.2	-35.5	-46.1	-55.1	-59.3	-64.8	-72.5	-78.7	-85.0	-90.7	-95.1	-96.8
Private	94.1	99.6	98.5	75.1	112.5	138.3	163.8	191.3	220.5	252.7	287.7	324.4	363.0
Exchange Losses	26.4	2.7	0.0	0.0	0.0	0.0	0.0	0.0	0.0	0.0	0.0	0.0	0.0
Other Net Assets	-71.7	-19.6	-14.4	-12.5	-13.4	-14.3	-15.3	-16.3	-17.3	-18.4	-19.6	-20.9	-22.3
Total Liabilities	126.7	167.3	164.2	175.4	197.0	221.8	245.7	272.1	300.5	332.1	367.7	406.7	450.6
M1	69.0	75.0	70.3	71.1	76.0	81.4	86.6	92.2	98.0	104.3	111.2	118.5	126.4
Currency	39.1	42.2	43.0	43.5	46.5	49.9	53.1	56.5	60.1	63.9	68.1	72.6	77.4
Demand Deposits	29.9	32.8	27.2	27.5	29.4	31.5	33.6	35.7	38.0	40.4	43.1	45.9	49.0
QM	48.7	78.8	84.0	95.7	113.2	133.0	151.9	173.1	195.9	221.5	250.5	282.4	318.5
Domestic Currency	44.0	44.0	41.7	51.5	62.1	74.0	85.7	98.5	112.5	127.9	145.1	164.0	185.0
Foreign Currency	4.7	34.8	42.4	44.2	51.1	59.0	66.3	74.6	83.4	93.5	105.3	118.4	133.5
Foreign Liabilities LT	9.0	13.5	9.9	8.6	7.8	7.5	7.1	6.8	6.5	6.3	6.1	5.9	5.7

High Case

Debt (Pipeline + New) (Mln. US$)	1988	1989	1990	1991	1992	1993	1994	1995	1996	1997	1998	1999	2000
Total Commitments	211.8	0.0	0.0	224.8	830.4	322.4	283.5	304.7	398.6	436.0	480.4	535.8	601.2
Refinancing	0.0	0.0	0.0	0.0	569.5	0.0	0.0	0.0	0.0	0.0	0.0	0.0	0.0
Other Multilateral and IDB	79.9	0.0	0.0	50.0	52.9	55.5	59.3	64.5	70.1	76.2	82.8	90.0	97.9
World Bank	0.0	0.0	0.0	5.0	30.0	100.0	53.4	58.0	63.1	68.6	74.6	81.0	88.1
Brazil (CACEX)	0.0	0.0	0.0	0.0	0.0	0.0	0.0	0.0	0.0	0.0	0.0	0.0	0.0
Bilateral	93.6	0.0	0.0	70.0	80.0	90.0	96.1	104.5	113.6	123.5	134.2	145.9	158.6
Suppliers + Export Credits	0.0	0.0	0.0	0.0	0.0	0.0	0.0	0.0	0.0	0.0	0.0	0.0	0.0
Financial Institutions	38.3	0.0	0.0	0.0	0.0	0.0	0.0	0.0	0.0	0.0	0.0	0.0	0.0
Private Non Guaranteed	0.0	0.0	0.0	0.0	0.0	0.0	0.0	0.0	0.0	0.0	0.0	0.0	0.0
IMF	0.0	0.0	0.0	0.0	0.0	0.0	0.0	0.0	0.0	0.0	0.0	0.0	0.0
Short Term	0.0	0.0	0.0	70.0	70.6	70.6	71.8	74.5	77.2	80.1	83.1	86.1	89.3
Unidentified Sources	0.0	0.0	0.0	29.8	27.4	6.3	2.8	3.2	74.6	87.7	105.8	132.7	167.5
Total Disbursements	141.8	628.8	80.0	240.7	813.4	244.6	228.8	264.1	354.0	381.9	422.4	475.5	536.2
Refinancing	0.0	0.0	0.0	0.0	569.5	0.0	0.0	0.0	0.0	0.0	0.0	0.0	0.0
Other Multilateral and IDB	31.4	18.0	15.0	24.6	33.6	44.2	42.0	54.4	59.3	63.6	65.8	71.6	77.5
World Bank	18.0	16.1	16.0	12.3	10.6	34.6	40.2	23.8	33.7	43.5	52.4	59.8	65.8
Brazil (CACEX)	0.0	426.6	0.0	0.0	0.0	0.0	0.0	0.0	0.0	0.0	0.0	0.0	0.0
Bilateral	40.4	125.9	22.0	70.7	81.4	75.2	71.9	99.3	109.1	107.0	115.3	125.3	136.2
Suppliers + Export Credits	14.8	41.2	25.0	25.6	15.2	10.1	0.0	2.6	0.0	0.0	0.0	0.0	0.0
Financial Institutions	37.2	0.9	2.0	7.6	5.2	3.6	0.0	6.3	0.0	0.0	0.0	0.0	0.0
Private Non Guaranteed	0.0	0.0	0.0	0.0	0.0	0.0	0.0	0.0	0.0	0.0	0.0	0.0	0.0
IMF	0.0	0.0	0.0	0.0	0.0	0.0	0.0	0.0	0.0	0.0	0.0	0.0	0.0
Short Term	-63.0	44.0	76.0	70.0	70.6	70.6	71.8	74.5	77.2	80.1	83.1	86.1	89.3
Unidentified Sources	0.0	0.0	0.0	29.8	27.4	6.3	2.8	3.2	74.6	87.7	105.8	132.7	167.5
Amortization Arrears	44.4	13.9	122.2	100.0	-425.0	0.0	0.0	0.0	0.0	0.0	0.0	0.0	0.0
Total Amortizations	180.8	449.8	533.9	203.4	185.7	179.2	157.3	165.8	270.0	289.1	317.0	349.5	403.8
Refinancing	0.0	0.0	0.0	0.0	0.0	0.0	0.0	0.0	81.4	81.4	81.4	81.4	81.4
Other Multilateral and IDB	13.0	22.4	21.0	27.1	29.3	30.8	31.8	40.4	49.3	60.6	67.2	77.9	89.6
World Bank	42.8	37.7	41.0	38.5	38.7	38.2	37.2	32.9	30.8	24.7	24.4	28.1	39.4
Brazil (CACEX)	0.0	0.0	426.6	0.0	0.0	0.0	0.0	0.0	0.0	6.4	6.4	6.4	6.4
Bilateral	41.1	59.5	28.0	44.1	44.9	45.7	42.2	59.4	75.5	91.6	113.2	130.6	149.6
Suppliers + Export Credits	40.3	5.7	16.0	42.5	38.1	33.0	29.6	17.9	15.6	8.5	8.5	8.5	8.5
Financial Institutions	43.6	323.1	0.0	49.6	32.9	29.3	14.8	8.3	6.0	3.9	3.9	3.9	3.9
Private Non Guaranteed	0.0	1.3	1.3	1.6	1.8	2.1	1.7	1.8	1.9	1.4	1.0	1.0	1.0
IMF	0.0	0.0	0.0	0.0	0.0	0.0	0.0	0.0	0.0	0.0	0.0	0.0	0.0
Unidentified Sources	0.0	0.0	0.0	0.0	0.0	0.0	0.0	5.0	9.5	10.6	11.1	11.6	24.0
High Case													

Debt (Pipeline + New) (Mln. US$) cont.	1988	1989	1990	1991	1992	1993	1994	1995	1996	1997	1998	1999	2000
Interest Arrears	15.7	-14.7	36.0	30.0	-144.5	0.0	0.0	0.0	0.0	0.0	0.0	0.0	0.0
Total Interest	114.6	91.4	90.0	108.2	107.1	127.4	123.9	124.0	114.7	114.8	120.0	131.3	143.7
Refinancing	0.0	0.0	0.0	0.0	22.8	43.3	40.4	39.9	33.8	28.6	23.4	18.2	13.0
Other Multilateral and IDB	20.6	20.5	22.0	21.3	21.0	21.3	21.9	22.5	22.3	22.4	22.4	24.0	25.3
World Bank	30.2	25.6	25.0	22.2	20.1	18.8	18.2	17.3	15.9	15.7	16.3	19.6	22.9
Brazil (CACEX)	0.0	14.8	0.0	0.0	0.0	0.0	0.0	0.0	0.0	-0.2	-0.3	-0.5	-0.6
Bilateral	17.7	9.2	10.0	16.3	17.1	18.3	19.3	20.5	19.6	19.5	22.4	24.9	27.0
Suppliers + Export Credits	25.0	7.0	17.0	12.6	11.2	8.7	6.2	3.9	2.3	1.4	0.7	0.7	0.7
Financial Institutions	19.4	2.5	1.0	10.9	7.2	4.5	2.7	1.6	1.0	0.5	0.2	0.2	0.2
Private Non Guaranteed	0.7	1.1	0.0	1.0	1.0	0.9	0.9	0.9	0.9	0.9	0.8	0.8	0.8
Interest on Arrears	0.0	9.8	14.0	22.8	0.0	0.0	0.0	0.0	0.0	0.0	0.0	0.0	0.0
IMF	0.0	0.0	0.0	0.0	0.0	0.0	0.0	0.0	0.0	0.0	0.0	0.0	0.0
Short Term	1.0	1.0	1.0	1.0	4.3	7.3	9.9	12.7	14.8	17.7	20.8	24.0	27.3
Unidentified Sources	0.0	0.0	0.0	0.0	2.4	4.3	4.5	4.6	4.1	8.3	13.2	19.3	27.0
Other Interest													
Accumulated Arrears	282.1	281.3	439.5	569.5	0.0	0.0	0.0	0.0	0.0	0.0	0.0	0.0	0.0
Total DOD	2354.9	2388.7	2127.5	2294.9	2353.0	2418.5	2489.9	2588.3	2672.3	2765.1	2870.5	2996.5	3129.0
Refinancing	0.0	0.0	0.0	0.0	569.5	569.5	569.5	569.5	488.1	406.8	325.4	244.1	162.7
Other Multilateral and IDB	419.4	409.7	415.0	412.5	416.9	430.3	440.5	454.4	464.5	467.5	466.1	459.7	447.6
World Bank	358.1	324.3	320.0	293.8	265.6	262.1	265.1	256.0	258.9	277.8	305.8	337.6	364.0
Brazil (CACEX)	0.0	426.6	0.0	0.0	0.0	0.0	0.0	0.0	0.0	-6.4	-12.8	-19.2	-25.6
Bilateral	478.2	450.4	470.0	496.7	533.1	562.5	592.3	632.1	665.7	681.0	683.2	677.8	664.5
Suppliers + Export credits	210.3	219.3	248.0	231.1	208.2	185.3	155.7	140.4	124.7	116.2	107.7	99.2	90.7
Financial Institutions	629.8	269.9	283.0	241.0	213.3	187.6	172.8	170.8	164.8	160.8	156.9	153.0	149.0
Private Non Guaranteed	28.0	28.0	19.0	17.4	15.6	13.5	11.8	10.0	8.1	6.7	5.7	4.6	3.6
IMF	0.0	0.0	0.0	0.0	0.0	0.0	0.0	0.0	0.0	0.0	0.0	0.0	0.0
Unidentified Sources	0.0	0.0	0.0	29.8	57.2	63.5	66.4	64.6	129.7	206.9	301.6	422.7	566.1
Short Term	231.2	260.5	372.5	472.5	398.6	469.1	541.0	615.4	692.7	772.8	855.8	942.0	1031.3
Interest Arrears	93.2	78.5	114.5	144.5	0.0	0.0	0.0	0.0	0.0	0.0	0.0	0.0	0.0
Other	138.0	182.0	258.0	328.0	398.6	469.1	541.0	615.4	692.7	772.8	855.8	942.0	1031.3
Adjustment in Debt Stock	-2.9	-3.7	-3.7	-3.7	-3.7	-3.7	-3.7	-3.7	-3.7	-3.7	-3.7	-3.7	-3.7
Net Disbursements:													
Other Multilateral and IDB	18.4	-4.4	-6.0	-2.5	4.3	13.4	10.2	14.0	10.1	3.0	-1.4	-6.4	-12.1
World Bank	-24.9	-21.6	0.0	-26.2	-28.1	-3.6	3.0	-9.1	2.9	18.8	28.1	31.7	26.4
Brazil (CACEX)	0.0	426.6	-6.6	0.0	0.0	0.0	0.0	0.0	0.0	-6.4	-6.4	-6.4	-6.4
Bilateral	-0.7	66.4	-6.0	26.7	36.4	29.5	29.7	39.8	33.6	15.3	2.1	-5.3	-13.4
Memorandum:													
Total Debt	2352.0	2385.0	2123.8	2291.2	2349.3	2414.8	2486.3	2584.6	2668.6	2761.4	2866.8	2992.8	3125.3
Debt Service/Exports (%)	37.3	41.2	43.3	23.6	14.0	13.1	10.9	10.0	12.0	11.4	11.2	11.1	11.4
Debt Service/GDP (%)	11.0	15.4	16.3	9.8	4.9	4.8	4.1	3.8	4.5	4.3	4.2	4.1	4.2
Interest/Exports (%)	11.2	5.0	7.0	7.4	5.1	5.4	4.8	4.3	3.6	3.2	3.1	3.0	3.0
Interest/GDP (%)	3.3	1.9	2.4	2.4	1.8	2.0	1.8	1.6	1.3	1.2	1.1	1.1	1.1
Debt/Exports (%)	227.4	173.5	142.3	153.0	112.6	103.0	96.4	89.7	83.7	78.2	73.2	69.1	65.1
Debt/GDP (%)	66.7	64.9	48.8	50.6	39.2	37.9	35.9	33.7	31.3	29.2	27.3	25.6	24.1

High Case

STATISTICAL APPENDIX

PARAGUAY

STATISTICAL APPENDIX

TABLE OF CONTENTS

Table 1.1 PARAGUAY - Total Population and Annual Growth Rates, 1960-2000

Table 1.2 PARAGUAY - Population Economically Active

Table 2.1 PARAGUAY - National Accounts by Sector of Origin, 1950-1990
(Current Guaranies, Billions)

Table 2.2 PARAGUAY - National Accounts by Sector of Origin, 1950-1990
(1982 Guaranies, Billions)

Table 2.3 PARAGUAY - National Accounts by Sector of Origin, 1950-1990
(Price Indices, 1982=100)

Table 2.4 PARAGUAY - National Accounts by Sector of Origin, 1950-1990
(Growth Rates)

Table 2.5 PARAGUAY -National Accounts by Expenditure, 1938-1990
(Current Guaranies, Billions)

Table 2.5 PARAGUAY - National Accounts by Expenditure, 1938-1990
(1982 Guaranies, Billions)

Table 2.7 PARAGUAY -National Accounts by Expenditure, 1938-1990
(Price Indices, 1982=100)

Table 2.8 PARAGUAY - National Accounts by Expenditure, 1938-1990
(Growth Rates, Percent)

Table 2.9 PARAGUAY - Harvested Area, Production Volume and Yield of Selected Agricultural Products, 1971-1990

Table 2.10 PARAGUAY - Domestic Prices for Selected Agricultural Goods, 1970-1989
(Producer Prices, Guaranies per kilo, Year Average)

Table 2.11 PARAGUAY - Industrial Sector, Value Added, by subsectors
(Current Guaranies)

Table 2.12 PARAGUAY - Industrial Sector, Value Added, by subsectors
(Constant 1982 Guaranies)

Table 2.13   PARAGUAY - Gross Domestic Investment, 1982-1989

Table 3.1   PARAGUAY - Balance of Payments, 1980-1990
        (US$ Millions)

Table 3.2   PARAGUAY - Composition of Registered Exports (FOB), 1976-1989
        (US$ Millions)

Table 3.3   PARAGUAY - Composition of Registered Imports (FOB), 1976-1989
        (US$ Millions)

Table 3.4   PARAGUAY - Average Market Exchange Rate, 1980-1990
        (Guaranies per US$)

Table 4.1   PARAGUAY - External Debt and Debt Service, 1975-1990
        (US$ Millions)

Table 5.1   PARAGUAY - Consolidated Public Sector Budget, 1970-1990
        (Current Guaranies, Millions)

Table 5.2   PARAGUAY - Central Administration Budget, 1970-1990
        (Current Guaranies, Millions)

Table 5.3   PARAGUAY - Decentralized Agencies Budget, 1970-1990
        (Current Guaranies, Millions)

Table 5.4   PARAGUAY - Municipalities Budget, 1970-1990
        (Current Guaranies, Millions)

Table 5.5   PARAGUAY - Public Enterprises Budget, 1970-1990
        (Current Guaranies, Millions)

Table 5.6   PARAGUAY - Executed Budget of Public Enterprises, 1980-1990
        (Current Guaranies, Millions)

Table 5.7   PARAGUAY - Executed Budget of Administracion Nacional de Electricidad (ANDE), 1980-1990
        (Current Guaranies, Millions)

Table 5.8   PARAGUAY - Executed Budget of Administracion Nacional de Telecomunicaciones (ANTELCO), 1980-1990
        (Current Guaranies, Millions)

Table 5.9   PARAGUAY - Executed Budget of Corporacion de Obras Sanitarias (CORPOSANA) 1980-1990
        (Current Guaranies, Millions)

Table 5.10   PARAGUAY - Executed Budget of Flota Mercante del Estado
             (FLOMERS), 1980-1990
                 (Current Guaranies, Millions)

Table 5.11   PARAGUAY - Executed Budget of Industria Nacional del Cemento
             (INC), 1980-1990
                 (Current Guaranies, Millions)

Table 5.12   PARAGUAY - Executed Budget of Lineas Aereas Paraguayas (LAP), 1980-1990
                 (Current Guaranies, Millions)

Table 5.13   PARAGUAY - Executed Budget of SIDEPAR and ACEPAR, 1980-1990
                 (Current Guaranies, Millions)

Table 5.14   PARAGUAY - Executed Budget of PETROPAR and APAL 1980-1990
                 (Current Guaranies, Millions)

Table 5.15   PARAGUAY - Executed Budget of Other Public Enterprises, 1980-1990
                 (Current Guaranies, Millions)

Table 5.16   PARAGUAY - Tax Revenue by Source, 1970-1990
                 (In millions of guaranies and as a percent of GDP)

Table 6.1   PARAGUAY - Accounts of the Consolidated Financial System, 1980-1990
                 (Current Guaranies Millions, End of Period)

Table 7.1   PARAGUAY - Consumer Price Index, 1972-1991
                 (1980 = 100)

Table 7.2   PARAGUAY - Wholesale Price Index, 1972-1990 (1980=100)
                 (1980 = 100)

Table 7.3   PARAGUAY - Wholesale Price Index, 1972-1991 (Growth Rates)
                 (Growth Rates)

Table 7.4   PARAGUAY - Wage Index, 1969-1990
                 (1980 = 100)

Table 1.1: Paraguay - Total Population and Annual Growth Rates, 1960 - 2000

Year	Total	Annual Growth Rate
1960	1,778,181	-
1965	2,018,892	2.6
1970	2,290,182	2.6
1971	2,358,973	3.0
1972	2,433,399	3.2
1973	2,513,165	3.3
1974	2,597,743	3.4
1975	2,686,457	3.4
1976	2,778,567	3.4
1977	2,857,555	3.3
1978	2,950,876	3.3
1979	3,047,245	3.3
1980	3,146,759	3.3
1981	3,250,256	3.3
1982	3,357,717	3.3
1983	3,468,078	3.3
1984	3,580,272	3.2
1985	3,693,233	3.2
1986	3,807,030	3.1
1987	3,922,374	3.0
1988	4,039,161	3.0
1989	4,157,287	2.9
1990	4,276,649	2.9
1995	4,853,044	2.6
2000	5,464,041	2.4

Source: Director General of Statistics and Census.

16-Oct-91

TABLE 1.2: PARAGUAY - Population Economically Active
(Thousands of persons)

Sector	1981	1982	1983	1984	1985	1986	1987	1988	1989	1990
Agriculture	417.1	418.3	422.4	470.3	493.2	463.3	498.5	560.2	599.8	609.5
Mining	2.2	2.3	2.4	2.5	2.6	2.7	2.8	2.9	3.0	3.1
Industry	129.0	126.2	125.7	142.4	149.4	149.0	152.1	156.9	162.2	164.5
Electricity, Water, and Sanitation	2.2	3.5	2.4	3.8	4.0	4.1	4.2	4.4	4.5	4.7
Construction	134.6	127.4	122.1	130.9	136.2	131.2	132.4	133.9	136.5	138.3
Commerce and Finance	150.3	147.2	144.1	160.1	166.6	169.5	174.6	181.0	186.9	192.1
Transport and Communication	29.2	29.2	29.3	33.0	34.4	36.9	39.4	42.9	45.8	48.1
Services	171.6	170.6	169.7	185.6	191.7	194.1	197.1	199.8	203.8	205.2
Others	33.6	33.9	33.0	36.9	38.3	41.0	43.7	46.8	54.0	59.7
Total employed	1069.8	1058.6	1051.1	1165.5	1216.4	1191.8	1244.8	1328.9	1396.5	1425.2
Unemployed	51.6	109.8	169.7	105.5	105.8	174.9	163.3	125.2	100.1	114.9
Pop. Econ. Active	1121.3	1168.5	1220.8	1271.0	1322.2	1366.7	1408.1	1454.1	1496.6	1540.0
Unemployment rate	4.6	9.4	13.9	8.3	8.0	12.8	11.6	8.6	6.7	7.5

Source: BCP and Director General of Statistics and Census

Table 2.1: PARAGUAY - NATIONAL ACCOUNTS BY SECTOR OF ORIGIN, 1950-1990
(Current Guaranies, Billions)

	1950	1951	1952	1953	1954	1955	1956	1957	1958	1959	1960	1961	1962	1963
Total Agriculture	0.63	1.14	2.17	3.96	4.82	6.41	8.21	9.44	9.63	10.96	13.01	14.70	16.82	18.48
Agriculture														
Livestock														
Forestry & Fishing														
Total Industry	0.3	0.5	0.8	1.4	1.8	2.3	3.0	4.2	4.8	5.4	6.6	7.5	8.2	8.7
Mining	0.00	0.00	0.00	0.00	0.00	0.01	0.01	0.03	0.04	0.03	0.06	0.06	0.05	0.07
Manufacturing	0.24	0.43	0.70	1.30	1.65	2.09	2.73	3.82	4.33	4.90	5.77	6.68	7.15	7.50
Construction	0.02	0.04	0.08	0.12	0.18	0.23	0.29	0.37	0.45	0.52	0.79	0.79	1.01	1.12
Total Basic Services	0.07	0.11	0.21	0.36	0.50	0.64	0.86	1.03	1.16	1.21	1.49	1.98	2.18	2.31
Electricity														
Water & Sanitation Services														
Transport & Communication	0.07	0.11	0.20	0.34	0.46	0.59	0.78	0.92	1.03	1.08	1.25	1.73	1.88	1.97
Total Services	0.57	0.91	1.61	3.04	4.23	5.83	7.53	9.85	11.76	12.66	14.46	16.37	18.24	18.90
Commerce & Finance	0.29	0.46	0.81	1.65	2.24	3.11	4.10	5.36	6.65	6.96	8.10	9.02	10.41	10.67
General Government	0.06	0.08	0.13	0.23	0.33	0.50	0.66	0.98	1.20	1.36	1.44	1.57	1.62	1.87
Housing	0.07	0.10	0.19	0.32	0.58	0.75	0.91	1.07	1.13	1.27	1.37	1.58	1.64	1.70
Other Services	0.2	0.3	0.5	0.8	1.1	1.5	1.9	2.4	2.8	3.1	3.6	4.2	4.6	4.6
GDP at Market Prices	1.53	2.64	4.77	8.79	11.38	15.20	19.63	24.55	27.37	30.28	35.58	40.58	45.45	48.37

Source: BCP and World Bank estimates.

Table 2.1: PARAGUAY - NATIONAL ACCOUNTS BY SECTOR OF ORIGIN, 1950-1990
(Current Guaranies, Billions)

	1964	1965	1966	1967	1968	1969	1970	1971	1972	1973	1974	1975	1976	1977
Total Agriculture	19.33	20.52	20.95	20.42	21.19	22.82	24.0	27.8	33.4	47.3	59.3	70.3	74.0	89.9
Agriculture							13.3	15.4	17.0	25.8	32.9	37.7	45.0	59.3
Livestock							7.3	8.8	12.4	16.4	19.6	23.8	21.3	21.8
Forestry & Fishing							3.4	3.6	4.0	5.0	6.9	8.7	7.6	8.8
Total Industry	9.4	10.1	11.0	12.1	12.3	13.4	14.7	16.3	18.4	23.7	36.0	37.3	43.8	56.2
Mining	0.10	0.10	0.19	0.16	0.06	0.07	0.1	0.2	0.2	0.2	0.3	0.4	0.5	0.7
Manufacturing	8.12	8.67	9.13	10.08	10.57	11.37	12.5	13.7	15.7	20.0	30.3	29.8	34.2	45.0
Construction	1.20	1.36	1.66	1.87	1.71	1.94	2.1	2.4	2.5	3.4	5.3	7.2	9.0	10.6
Total Basic Services	2.61	2.77	2.88	3.15	3.26	3.55	3.8	4.3	5.1	6.3	8.2	10.3	12.4	14.9
Electricity							0.7	0.9	1.1	1.6	1.7	2.3	3.2	4.0
Water & Sanitation Services							0.1	0.2	0.2	0.3	0.3	0.4	0.5	0.7
Transport & Communication	2.26	2.40	2.49	2.67	2.76	2.93	3.0	3.2	3.8	4.3	6.1	7.6	8.7	10.3
Total Services	20.08	22.48	23.88	26.40	28.44	30.35	32.45	35.33	39.97	48.22	64.52	72.53	83.9	102.6
Commerce & Finance	11.31	12.74	13.35	15.10	16.00	17.23	18.3	20.2	22.3	28.9	39.9	43.6	51.5	66.0
General Government	1.98	2.15	2.43	2.66	3.18	3.45	3.9	4.2	4.6	4.8	5.3	6.5	7.6	10.3
Housing	1.76	1.88	1.97	2.01	2.06	2.21	2.3	2.3	2.6	2.8	4.1	5.0	5.6	6.1
Other Services	5.0	5.7	6.1	6.6	7.2	7.4	7.9	8.6	10.5	11.7	15.3	17.4	19.2	20.2
GDP at Market Prices	51.45	55.89	58.70	62.08	65.22	70.09	74.9	83.7	96.9	125.4	168.0	190.4	214.1	263.6

Source: BCP and World Bank estimates.

Table 2.1: PARAGUAY - NATIONAL ACCOUNTS BY SECTOR OF ORIGIN, 1950-1990
(Current Guaranies, Billions)

	1978	1979	1980	1981	1982	1983	1984	1985	1986	1987	1988	1989	1990
Total Agriculture	103.4	135.2	165.1	196.8	190.6	210.6	307.1	403.3	498.9	681.9	983.3	1361.8	1798.7
Agriculture	63.2	84.2	101.2	120.1	114.7	127.3	191.1	253.6	288.0	396.9	660.0	906.6	1178.8
Livestock	30.1	38.5	46.7	55.4	56.5	62.7	81.9	104.5	140.2	187.7	221.4	311.8	435.1
Forestry & Fishing	10.0	12.5	17.2	21.3	19.5	20.6	34.2	45.2	70.7	97.3	101.8	143.4	184.9
Total Industry	70.7	94.3	128.9	168.1	173.7	192.8	243.2	314.7	414.5	561.6	735.2	1066.0	1494.8
Mining	0.8	1.4	2.3	2.9	3.1	3.5	4.4	5.7	8.4	11.5	15.7	21.2	23.1
Manufacturing	54.4	69.6	92.3	118.5	121.0	134.3	172.0	226.1	296.0	404.1	556.1	785.6	1118.7
Construction	15.5	23.2	34.3	46.7	49.5	55.0	66.9	82.9	110.1	146.0	163.4	259.3	353.1
Total Basic Services	19.0	25.2	36.7	44.3	49.2	55.2	70.1	88.9	124.9	171.9	215.8	286.2	411.7
Electricity	5.1	6.8	11.2	13.1	15.8	17.5	22.1	25.8	37.7	52.9	72.1	96.6	140.3
Water & Sanitation Services	0.9	1.1	1.7	2.1	2.3	3.1	3.9	5.1	7.1	9.7	11.7	15.2	21.5
Transport & Communication	13.0	17.4	23.8	29.0	31.1	34.5	44.1	58.0	80.1	109.3	132.1	174.4	249.9
Total Services	129.5	175.8	229.7	299.4	323.5	358.6	450.0	587.1	795.5	1078.1	1384.8	1894.4	2769.2
Commerce & Finance	84.0	112.7	144.9	188.4	196.2	217.2	272.6	360.4	489.8	662.2	914.0	1262.7	1880.7
General Government	12.7	14.6	19.1	26.7	32.9	36.5	45.2	58.3	78.3	104.8	120.8	172.5	214.1
Housing	7.5	11.2	15.0	20.1	22.5	25.0	31.0	39.2	52.6	70.4	77.7	91.9	123.3
Other Services	25.3	37.4	50.7	64.3	72.0	79.9	101.2	129.2	174.9	240.8	272.4	367.3	551.1
GDP at Market Prices	322.5	430.5	560.5	708.7	737.0	817.1	1070.4	1393.9	1833.8	2493.6	3319.1	4608.4	6474.4

Source: BCP and World Bank estimates.

Table 2.2: PARAGUAY - NATIONAL ACCOUNTS BY SECTOR OF ORIGIN, 1950-1990
(1982 Guaranies, Billions)

	1950	1951	1952	1953	1954	1955	1956	1957	1958	1959	1960	1961	1962	1963
Total Agriculture	55.6	56.0	57.3	57.7	57.3	61.2	62.9	64.0	66.1	68.2	66.6	71.7	73.8	76.4
Agriculture														
Livestock														
Forestry & Fishing														
Total Industry	25.4	27.4	27.3	27.9	28.1	29.4	28.8	29.1	31.1	33.5	32.7	34.9	38.6	40.7
Mining	0.0	0.0	0.0	0.0	0.0	0.0	0.0	0.1	0.1	0.1	0.2	0.2	0.1	0.3
Manufacturing	23.8	25.5	25.2	26.2	26.2	27.2	26.4	26.4	28.2	30.0	28.8	31.2	34.7	36.0
Construction	1.7	1.8	2.1	1.7	1.9	2.2	2.3	2.6	2.8	3.4	3.8	3.5	3.8	4.4
Total Basic Services	7.8	6.9	7.5	7.7	7.8	8.2	8.5	8.7	8.6	8.5	9.2	11.2	9.7	10.1
Electricity														
Water & Sanitation Services														
Transport & Communication	7.5	6.5	7.1	7.3	7.3	7.7	8.0	8.1	7.9	7.7	8.1	10.1	8.5	8.8
Total Services	50.3	49.9	47.8	54.0	58.3	64.0	66.5	74.5	81.5	77.0	77.1	80.7	82.9	86.4
Commerce & Finance	34.6	32.0	30.4	35.7	39.1	44.2	45.2	50.8	56.5	51.0	48.9	52.0	52.6	53.0
General Government	5.8	6.2	6.5	6.5	6.8	6.8	7.2	7.6	7.7	7.8	8.0	8.0	8.0	9.1
Housing	6.9	7.1	7.3	7.5	7.7	7.9	8.1	8.4	8.6	8.9	9.2	9.4	9.7	10.0
Other Services	3.0	4.6	3.5	4.2	4.7	5.1	6.0	7.8	8.7	9.4	11.1	11.3	12.6	14.4
GDP at Market Prices	139.1	140.2	139.9	147.2	151.4	162.8	166.7	176.4	187.3	187.1	185.6	198.4	205.0	213.6

Source: BCP and World Bank estimates.

Table 2.2: PARAGUAY - NATIONAL ACCOUNTS BY SECTOR OF ORIGIN, 1950-1990
(1982 Guaranies, Billions)

	1964	1965	1966	1967	1968	1969	1970	1971	1972	1973	1974	1975	1976	1977
Total Agriculture	78.5	82.7	81.6	86.5	85.7	87.8	89.8	95.5	100.7	106.8	116.8	128.2	133.2	142.0
Agriculture														80.6
Livestock														47.8
Forestry & Fishing														13.6
Total Industry	42.9	44.6	47.1	52.9	54.1	57.1	61.6	63.8	70.2	76.7	83.1	84.3	90.8	108.8
Mining	0.4	0.4	0.6	0.5	0.2	0.2	0.2	0.5	0.5	0.5	0.6	0.7	1.0	1.3
Manufacturing	38.1	39.4	40.7	45.8	47.9	50.2	54.3	55.2	61.3	66.4	71.3	70.0	73.8	86.6
Construction	4.5	4.9	5.8	6.6	6.0	6.7	7.1	8.2	8.4	9.8	11.1	13.5	15.9	20.9
Total Basic Services	11.2	11.5	11.6	12.4	12.8	13.5	14.5	15.5	17.1	19.4	21.5	24.8	27.5	30.1
Electricity														7.2
Water & Sanitation Services														1.2
Transport & Communication	9.9	10.1	10.2	10.8	11.1	11.5	11.7	12.1	13.0	14.4	16.4	18.7	20.0	21.6
Total Services	89.9	96.5	98.5	107.9	116.3	122.3	129.4	137.4	145.0	155.2	167.1	179.2	195.2	214.7
Commerce & Finance	55.1	59.7	60.8	67.9	71.5	75.2	78.3	83.1	85.7	93.0	101.2	105.6	116.3	130.2
General Government	9.4	9.8	10.8	11.7	13.7	14.7	17.0	17.1	17.2	15.9	15.4	17.9	19.1	20.3
Housing	10.3	9.2	9.5	9.7	10.0	10.3	10.7	11.1	11.4	12.1	12.7	13.7	14.8	16.0
Other Services	15.2	17.7	17.4	18.6	21.2	22.0	23.4	26.1	30.7	34.2	37.8	42.0	45.0	48.2
GDP at Market Prices	222.6	235.3	238.8	259.8	268.9	280.7	295.3	312.2	333.1	358.0	388.5	416.5	446.7	495.5

Source: BCP and World Bank estimates.

Table 2.2: PARAGUAY - NATIONAL ACCOUNTS BY SECTOR OF ORIGIN, 1950-1990
(1982 Guaranies, Billions)

	1978	1979	1980	1981	1982	1983	1984	1985	1986	1987	1988	1989	1990
Total Agriculture	149.1	159.0	172.5	189.9	190.6	186.0	197.0	206.0	193.5	207.0	232.0	249.8	255.4
Agriculture	84.5	90.4	99.3	113.9	114.7	111.4	119.7	126.9	110.9	121.6	143.5	157.6	159.1
Livestock	49.7	51.7	53.8	55.4	56.5	55.5	57.8	59.4	60.6	62.1	63.9	66.6	69.8
Forestry & Fishing	14.8	16.9	19.4	20.5	19.5	19.1	19.6	19.7	22.0	23.3	24.6	25.6	26.5
Total Industry	125.8	144.3	168.3	181.4	173.7	165.5	169.6	175.4	174.4	179.9	189.0	198.5	201.8
Mining	1.5	2.1	2.7	3.1	3.1	2.9	2.9	3.1	3.4	3.6	3.9	4.1	4.3
Manufacturing	96.7	106.3	120.4	125.6	121.0	115.9	121.1	127.1	125.3	129.7	137.3	145.4	149.0
Construction	27.6	35.8	45.2	52.7	49.5	46.7	45.6	45.1	45.6	46.5	47.7	48.9	48.5
Total Basic Services	33.9	38.5	43.6	45.2	49.2	48.5	50.0	52.7	56.4	59.7	64.0	67.3	72.5
Electricity	8.4	10.1	12.1	12.6	15.8	15.0	15.3	16.3	18.1	19.5	21.4	22.7	26.0
Water & Sanitation Services	1.5	1.7	1.9	2.1	2.3	2.8	2.8	3.0	3.2	3.3	3.5	3.6	4.0
Transport & Communication	24.0	26.7	29.6	30.5	31.1	30.7	31.9	33.5	35.1	36.9	39.1	41.0	42.5
Total Services	242.9	272.6	300.3	327.9	323.5	314.9	320.3	332.1	342.0	352.8	365.3	383.8	397.6
Commerce & Finance	148.8	167.4	185.0	200.6	196.2	190.2	193.6	202.8	209.4	216.8	225.6	236.1	244.7
General Government	21.9	24.0	25.7	31.6	32.9	32.2	33.0	33.9	34.6	35.6	35.9	40.6	41.8
Housing	17.8	19.7	21.5	23.0	22.5	21.4	21.4	21.7	22.1	22.5	23.0	23.5	24.1
Other Services	54.4	61.4	68.1	72.7	72.0	71.1	72.2	73.7	75.9	77.8	80.8	83.6	87.0
GDP at Market Prices	551.7	614.4	684.7	744.3	737.0	714.9	736.9	766.2	766.2	799.4	850.2	899.5	927.3

Source: BCP and World Bank estimates.

Table 2.3: PARAGUAY - NATIONAL ACCOUNTS BY SECTOR OF ORIGIN, 1950-1990
(Price Indices, 1982=100)

	1950	1951	1952	1953	1954	1955	1956	1957	1958	1959	1960	1961	1962	1963
Total Agriculture	1.1	2.0	3.8	6.9	8.4	10.5	13.1	14.7	14.6	16.1	19.5	20.5	22.8	24.2
Agriculture														
Livestock														
Forestry & Fishing														
Total Industry	1.0	1.8	2.8	5.1	6.5	7.9	10.5	14.5	15.5	16.3	20.2	21.6	21.3	21.4
Mining	2.4	2.0	9.6	16.8	16.5	17.0	20.2	23.3	25.5	27.5	29.4	33.5	33.5	25.3
Manufacturing	1.0	1.7	2.8	5.0	6.3	7.7	10.3	14.5	15.4	16.4	20.1	21.4	20.6	20.9
Construction	1.4	2.5	3.5	7.4	9.7	10.5	12.5	14.4	16.3	15.3	21.0	23.0	27.0	25.2
Total Basic Services	0.9	1.6	2.9	4.7	6.4	7.8	10.1	11.8	13.5	14.3	16.2	17.7	22.5	22.9
Electricity														
Water & Sanitation Services														
Transport & Communication	0.9	1.6	2.8	4.6	6.3	7.6	9.7	11.3	13.0	14.1	15.4	17.2	22.0	22.4
Total Services	1.1	1.8	3.4	5.6	7.3	9.1	11.3	13.2	14.4	16.4	18.8	20.3	22.0	21.9
Commerce & Finance	0.8	1.4	2.7	4.6	5.7	7.0	9.1	10.6	11.8	13.6	16.6	17.4	19.8	20.1
General Government	1.0	1.3	2.0	3.5	4.9	7.4	9.2	13.0	15.6	17.5	18.0	19.7	20.3	20.7
Housing	1.0	1.5	2.7	4.2	7.6	9.5	11.1	12.8	13.1	14.2	14.9	16.7	16.9	17.0
Other Services	5.2	5.7	13.3	20.0	22.8	28.8	31.1	31.5	31.9	32.8	32.2	37.3	36.3	32.3
GDP at Market Prices	1.1	1.9	3.4	6.0	7.5	9.3	11.8	13.9	14.6	16.2	19.2	20.4	22.2	22.6

Source: BCP and World Bank estimates.

Table 2.3: PARAGUAY - NATIONAL ACCOUNTS BY SECTOR OF ORIGIN, 1950-1990
(Price Indices, 1982 = 100)

	1964	1965	1966	1967	1968	1969	1970	1971	1972	1973	1974	1975	1976	1977
Total Agriculture	24.6	24.8	25.7	23.6	24.7	26.0	26.7	29.1	33.1	44.3	50.8	54.8	55.5	63.3
Agriculture														73.6
Livestock														45.5
Forestry & Fishing														65.1
Total Industry	22.0	22.7	23.3	22.9	22.8	23.4	23.8	25.6	26.3	30.9	43.3	44.3	48.2	51.7
Mining	27.9	29.0	33.0	34.6	33.5	36.3	37.9	39.0	39.2	38.6	49.0	49.9	50.8	53.7
Manufacturing	21.3	22.0	22.4	22.0	22.1	22.6	23.0	24.9	25.6	30.2	42.5	42.5	46.4	51.9
Construction	27.0	27.8	28.5	28.1	28.5	29.1	29.1	29.7	30.3	35.1	48.0	53.1	56.7	50.6
Total Basic Services	23.3	24.2	24.9	25.3	25.5	26.2	26.1	27.5	29.8	32.3	38.2	41.6	45.2	49.4
Electricity														54.7
Water & Sanitation Services														52.7
Transport & Communication	22.8	23.7	24.4	24.7	24.9	25.5	25.2	26.5	28.9	30.1	37.4	40.6	43.6	47.5
Total Services	22.3	23.3	24.3	24.5	24.4	24.8	25.1	25.7	27.6	31.1	38.6	40.5	43.0	47.8
Commerce & Finance	20.5	21.3	22.0	22.2	22.4	22.9	23.4	24.3	26.0	31.1	39.4	41.3	44.3	50.7
General Government	21.1	21.9	22.5	22.8	23.1	23.5	23.3	24.4	26.7	30.1	34.4	36.2	39.9	50.7
Housing	17.1	20.4	20.8	20.7	20.6	21.4	21.3	21.2	22.8	23.5	32.3	36.7	37.7	38.0
Other Services	33.1	32.2	35.2	35.7	34.1	33.9	33.9	33.0	34.2	34.2	40.4	41.5	42.6	42.0
GDP at Market Prices	23.1	23.8	24.6	23.9	24.3	25.0	25.4	26.8	29.1	35.0	43.3	45.7	47.9	53.2

Source: BCP and World Bank estimates.

Table 2.3: PARAGUAY - NATIONAL ACCOUNTS BY SECTOR OF ORIGIN, 1950-1990
(Price Indices, 1982 = 100)

	1978	1979	1980	1981	1982	1983	1984	1985	1986	1987	1988	1989	1990
Total Agriculture	69.4	85.0	95.7	103.6	100.0	113.2	155.9	195.7	257.9	329.4	423.8	545.1	704.3
Agriculture	74.8	93.1	102.0	105.4	100.0	114.2	159.7	199.9	259.8	326.3	459.8	575.2	741.0
Livestock	60.6	74.4	86.7	100.0	100.0	113.0	141.7	175.8	231.3	302.2	346.7	467.9	622.9
Forestry & Fishing	67.7	74.0	88.7	103.7	100.0	108.0	174.7	229.0	321.8	418.5	414.1	560.6	698.7
Total Industry	56.2	65.3	76.6	92.7	100.0	116.5	143.4	179.5	237.7	312.2	389.0	537.0	740.6
Mining	53.4	68.3	85.6	95.5	100.0	119.7	148.2	184.6	242.7	315.6	400.7	510.2	536.4
Manufacturing	56.3	65.5	76.7	94.3	100.0	115.9	142.1	177.9	236.2	311.5	405.0	540.3	750.6
Construction	56.1	64.7	76.0	88.7	100.0	117.7	146.6	183.6	241.5	314.0	342.2	529.8	728.2
Total Basic Services	56.0	65.5	84.2	98.0	100.0	113.7	140.2	168.6	221.6	287.9	337.4	425.0	568.0
Electricity	60.6	67.2	92.6	104.2	100.0	116.7	144.2	158.6	208.6	271.2	337.4	425.0	540.6
Water & Sanitation Services	58.5	64.8	89.2	100.0	100.0	113.0	139.6	170.9	225.1	292.7	337.4	425.0	540.6
Transport & Communication	54.2	64.9	80.5	95.3	100.0	112.3	138.3	173.2	227.9	296.2	337.4	425.0	587.3
Total Services	53.3	64.5	76.5	91.3	100.0	113.9	140.5	176.8	232.6	305.6	379.1	493.5	696.4
Commerce & Finance	56.4	67.3	78.3	93.9	100.0	114.2	140.8	177.7	233.9	305.5	405.1	534.7	768.5
General Government	58.1	60.7	74.3	84.4	100.0	113.4	137.2	171.8	226.1	294.3	337.0	425.0	512.5
Housing	42.0	57.0	69.8	87.5	100.0	116.4	144.4	180.8	237.8	312.2	337.8	390.2	511.7
Other Services	46.5	60.8	74.4	88.4	100.0	112.4	140.1	175.3	230.6	309.4	337.2	439.5	633.4
GDP at Market Prices	58.5	70.1	81.9	95.2	100.0	114.3	145.3	181.9	239.3	311.9	390.4	512.3	698.2

Source: BCP and World Bank estimates.

Table 2.4: PARAGUAY - NATIONAL ACCOUNTS BY SECTOR OF ORIGIN, 1950-1990
(Growth Rates)

	1950	1951	1952	1953	1954	1955	1956	1957	1958	1959	1960	1961	1962	1963
Total Agriculture		0.9	2.3	0.6	-0.6	6.8	2.8	1.8	3.2	3.2	-2.3	7.7	2.9	3.5
Agriculture														
Livestock														
Forestry & Fishing														
Total Industry		7.5	-0.2	2.2	0.7	4.7	-2.1	1.1	6.9	7.7	-2.2	6.6	10.7	5.4
Mining		50.3	-17.7	-20.1	5.4	110.8	32.1	195.7	-1.9	-30.9	103.3	-14.9	-18.2	110.1
Manufacturing		7.5	-1.5	4.2	-0.2	3.9	-2.8	-0.2	6.8	6.4	-4.0	8.6	11.1	3.6
Construction		7.3	18.7	-21.5	14.1	14.7	5.4	11.9	8.2	22.3	10.2	-8.0	8.7	18.3
Total Basic Services		-11.8	9.1	2.0	1.5	5.8	3.8	2.6	-1.3	-2.0	9.0	21.8	-13.5	3.7
Electricity														
Water & Sanitation Services														
Transport & Communication		-12.8	9.6	1.8	1.0	5.4	3.3	1.9	-2.7	-2.8	5.6	24.1	-15.5	3.3
Total Services		-0.7	-4.2	12.9	8.0	9.8	4.0	12.0	9.4	-5.5	0.1	4.6	2.8	4.3
Commerce & Finance		-7.4	-5.0	17.4	9.5	13.0	2.2	12.4	11.2	-9.7	-4.1	6.2	1.3	0.7
General Government		6.0	5.9	-0.1	3.8	-0.2	6.4	5.1	1.8	0.6	2.6	0.3	0.1	13.5
Housing		2.9	2.9	2.9	2.9	2.9	3.0	3.0	2.9	3.1	2.9	3.0	3.0	2.8
Other Services		55.2	-23.3	19.1	11.1	8.3	17.6	29.7	12.3	7.7	18.0	2.0	11.4	14.5
GDP at Market Prices		0.8	-0.2	5.2	2.9	7.5	2.4	5.8	6.2	-0.1	-0.8	6.9	3.3	4.2

Source: BCP and World Bank estimates.

Table 2.4: PARAGUAY - NATIONAL ACCOUNTS BY SECTOR OF ORIGIN, 1950-1990
(Growth Rates)

	1964	1965	1966	1967	1968	1969	1970	1971	1972	1973	1974	1975	1976	1977
Total Agriculture	2.7	5.4	-1.3	6.0	-1.0	2.4	2.3	6.2	5.5	6.0	9.4	9.7	3.9	6.6
Agriculture														
Livestock														
Forestry & Fishing														
Total Industry	5.5	3.9	5.7	12.2	2.2	5.6	7.9	3.6	9.9	9.3	8.3	1.4	7.8	19.8
Mining	24.7	-1.4	63.8	-19.4	-63.9	11.1	15.3	116.4	13.9	-1.3	14.1	20.2	42.3	22.7
Manufacturing	6.0	3.3	3.4	12.5	4.6	4.9	8.0	1.7	11.0	8.4	7.4	-1.8	5.4	17.4
Construction	0.5	9.4	19.7	13.6	-9.4	10.8	7.2	14.3	2.6	16.5	14.1	21.2	18.1	31.1
Total Basic Services	11.6	2.2	0.8	7.5	2.9	5.9	7.0	7.1	10.3	13.1	11.1	15.4	10.8	9.3
Electricity														
Water & Sanitation Services														
Transport & Communication	12.4	2.4	0.7	5.8	3.0	3.3	1.9	3.5	7.7	10.4	14.1	13.8	6.7	8.3
Total Services	4.0	7.3	2.0	9.6	7.8	5.1	5.8	6.2	5.6	7.0	7.7	7.3	8.9	10.0
Commerce & Finance	3.9	8.5	1.7	11.8	5.2	5.3	4.0	6.2	3.1	8.5	8.8	4.4	10.1	12.0
General Government	3.4	4.8	9.9	8.1	17.0	7.6	15.3	0.8	0.8	-7.7	-3.3	16.6	6.7	6.0
Housing	2.9	-10.6	3.0	2.9	3.0	3.2	3.4	3.3	3.3	6.0	5.3	7.2	8.1	8.3
Other Services	5.7	16.6	-1.7	6.6	13.9	3.9	6.6	11.3	17.6	11.4	10.6	11.1	7.2	7.0
GDP at Market Prices	4.2	5.7	1.5	8.8	3.5	4.4	5.2	5.7	6.7	7.5	8.5	7.2	7.3	10.9

Source: BCP and World Bank estimates.

Table 2.4: PARAGUAY - NATIONAL ACCOUNTS BY SECTOR OF ORIGIN, 1950-1990
(Growth Rates)

	1978	1979	1980	1981	1982	1983	1984	1985	1986	1987	1988	1989	1990
Total Agriculture	5.0	6.6	8.5	10.1	0.4	-2.4	5.9	4.6	-6.1	7.0	12.1	7.7	2.2
Agriculture	4.9	7.0	9.8	14.8	0.6	-2.8	7.4	6.0	-12.6	9.7	18.0	9.8	0.9
Livestock	4.0	4.0	4.0	3.0	2.0	-1.8	4.1	2.9	2.0	2.5	2.8	4.3	4.8
Forestry & Fishing	9.3	13.6	15.3	5.6	-5.2	-1.9	2.5	1.0	11.2	5.8	5.8	4.0	3.4
Total Industry	15.6	14.7	16.6	7.8	-4.3	-4.7	2.5	3.4	-0.6	3.2	5.0	5.0	1.7
Mining	16.5	42.4	26.0	15.0	2.3	-7.3	1.0	4.5	11.9	6.0	7.5	5.8	3.7
Manufacturing	11.6	9.9	13.3	4.3	-3.7	-4.2	4.5	5.0	-1.4	3.5	5.8	5.9	2.5
Construction	32.0	30.0	26.0	16.7	-6.0	-5.7	-2.4	-1.0	1.0	2.0	2.6	2.5	-0.9
Total Basic Services	12.7	13.7	13.1	3.8	8.9	-1.4	3.1	5.4	6.9	6.0	7.1	5.3	7.6
Electricity	16.1	20.1	20.4	4.0	25.0	-4.8	2.2	5.9	11.1	8.0	9.6	6.4	14.2
Water & Sanitation Services	23.3	12.2	10.0	12.4	10.3	18.1	2.0	6.1	5.5	5.0	4.2	3.7	10.9
Transport & Communication	11.0	11.5	10.5	3.1	2.1	-1.2	3.6	5.1	5.0	5.0	6.1	4.8	3.7
Total Services	13.2	12.2	10.2	9.2	-1.3	-2.7	1.7	3.7	3.0	3.1	3.5	5.1	3.6
Commerce & Finance	14.3	12.5	10.5	8.4	-2.2	-3.1	1.8	4.7	3.3	3.5	4.1	4.7	3.6
General Government	7.9	9.9	7.0	22.8	4.0	-2.1	2.4	3.0	2.0	2.9	0.7	13.2	2.9
Housing	11.5	10.5	9.0	6.9	-2.0	-4.7	0.0	1.0	2.0	2.0	2.0	2.4	2.4
Other Services	12.9	12.9	10.9	6.8	-1.0	-1.2	1.6	2.0	2.9	2.6	3.8	3.4	4.1
GDP at Market Prices	11.4	11.4	11.4	8.7	-1.0	-3.0	3.1	4.0	0.0	4.3	6.4	5.8	3.1

Source: BCP and World Bank estimates.

Table 2.5: PARAGUAY - NATIONAL ACCOUNTS BY EXPENDITURE, 1938-1990
(Current Guaranies, Billions)

	1938	1939	1940	1941	1942	1943	1944	1945	1946	1947	1948	1949	1950	1951	1952	1953	1954	1955
Consumption													1.3	2.3	4.5	8.1	10.8	14.0
Private													1.23	2.16	4.13	7.43	9.92	13.00
Public													0.11	0.16	0.36	0.65	0.86	0.98
Investment													0.11	0.19	0.32	0.59	0.72	1.13
Fixed													0.10	0.16	0.30	0.56	0.78	1.04
Private																		
Public																		
Change in Inventories													0.02	0.03	0.02	0.03	-0.05	0.09
Total Expenditures													1.44	2.51	4.82	8.66	11.51	15.11
Imports G&NFS													0.25	0.54	1.27	1.83	2.72	2.80
Exports G&NFS													0.34	0.67	1.23	1.95	2.59	2.89
GDP													1.53	2.64	4.77	8.79	11.38	15.20
Net Factor Income																		
Gross National Product																		
Memo Items:																		
Resource Balance													0.1	0.1	0.0	0.1	-0.1	0.1
Expend. on Domestic Goods a/													1.2	2.0	3.5	6.8	8.8	12.3

a/ Expenditures minus imports.
Source: BCP and World Bank estimate.
16-Oct-91

Table 2.5: PARAGUAY - NATIONAL ACCOUNTS BY EXPENDITURE, 1938-1990

(Current Guaranies, Billions)

	1956	1957	1958	1959	1960	1961	1962	1963	1964	1965	1966	1967	1968	1969	1970	1971	1972	1973
Consumption	18.3	23.5	25.5	27.9	32.5	36.7	40.7	43.7	46.4	47.9	51.1	53.9	57.8	62.1	64.8	73.7	82.4	100.7
Private	16.97	21.69	23.29	25.47	29.92	33.74	37.48	40.29	42.92	44.11	46.53	49.10	52.33	55.76	58.04	66.58	74.61	92.54
Public	1.28	1.79	2.18	2.47	2.61	2.93	3.20	3.36	3.50	3.79	4.59	4.83	5.43	6.33	6.75	7.11	7.77	8.17
Investment	1.35	1.82	3.20	2.88	4.40	5.23	5.70	5.40	6.10	8.43	9.28	10.27	10.35	11.24	11.03	12.20	14.59	23.86
Fixed	1.22	1.73	3.12	2.93	4.28	4.73	5.24	4.94	5.64	7.99	9.08	10.09	10.04	10.81	10.88	11.80	13.27	20.41
Private																		
Public																		
Change in Inventories	0.12	0.09	0.08	-0.05	0.12	0.50	0.46	0.46	0.46	0.44	0.20	0.17	0.31	0.43	0.15	0.40	1.32	3.45
Total Expenditures	19.60	25.30	28.67	30.82	36.93	41.91	46.38	49.06	52.52	56.34	60.40	64.19	68.10	73.33	75.82	85.89	96.97	124.57
Imports G&NFS	3.44	5.25	5.88	5.58	6.59	7.39	6.67	6.28	7.38	8.80	9.48	9.70	10.94	12.76	12.08	13.35	13.41	17.90
Exports G&NFS	3.47	4.50	4.58	5.04	5.24	6.05	5.74	5.59	6.31	8.36	7.78	7.58	8.06	9.52	11.18	11.20	13.34	18.77
GDP	19.63	24.55	27.37	30.28	35.58	40.58	45.45	48.37	51.45	55.89	58.70	62.08	65.22	70.09	74.92	83.74	96.90	125.44
Net Factor Income																		
Gross National Product																		
Memo Items:																		
Resource Balance	0.0	-0.7	-1.3	-0.5	-1.3	-1.3	-0.9	-0.7	-1.1	-0.4	-1.7	-2.1	-2.9	-3.2	-0.9	-2.2	-0.1	0.9
Expend. on Domestic Goods a/	16.2	20.1	22.8	25.2	30.3	34.5	39.7	42.8	45.1	47.5	50.9	54.5	57.2	60.6	63.7	72.5	83.6	106.7

a/ Expenditures minus imports.
Source: BCP and World Bank estimate
16-Oct-91

Table 2.5: PARAGUAY - NATIONAL ACCOUNTS BY EXPENDITURE, 1938-1990

(Current Guaranies, Billions)

	1974	1975	1976	1977	1978	1979	1980	1981	1982	1983	1984	1985	1986	1987	1988	1989	1990
Consumption	136.6	154.9	168.6	206.4	246.7	331.2	434.0	552.7	604.3	700.2	891.2	1157.1	1556.6	2059.2	2795.8	3803.7	5243.7
Private	127.33	142.97	155.2	190.1	225.2	306.5	399.3	504.1	552.0	642.2	822.0	1066.9	1434.8	1882.6	2586.7	3497.0	4841.8
Public	9.23	11.97	13.4	16.4	21.5	24.7	34.7	48.6	52.3	58.0	69.3	90.2	121.8	176.6	209.1	306.7	401.9
Investment	35.27	45.89	52.7	65.1	87.7	123.0	161.2	204.3	188.9	175.2	245.5	306.5	458.9	625.8	808.7	1098.8	1480.4
Fixed	30.90	39.54	48.7	62.9	81.3	116.1	152.7	194.2	176.9	164.5	231.2	288.0	431.8	591.4	768.2	1045.6	1425.4
Private			29.0	40.1	57.6	89.1	125.3	159.9	143.7	130.9	172.9	202.1	306.5	414.9	549.8	880.3	1200.1
Public			19.7	22.8	23.7	27.1	27.4	34.3	33.2	33.6	58.3	85.8	125.3	176.5	218.4	165.3	225.3
Change in Inventories	4.37	6.35	4.0	2.2	6.5	6.8	8.6	10.1	12.0	10.7	14.3	18.5	27.1	34.4	40.6	53.2	55.0
Total Expenditures	171.83	200.83	221.3	271.5	334.5	454.2	595.2	757.0	793.2	875.5	1136.7	1463.6	2015.5	2685.0	3604.6	4902.5	6724.1
Imports G&NFS	29.87	35.55	38.6	59.2	71.3	92.8	112.4	127.4	145.6	127.4	236.3	424.7	580.7	896.1	1224.4	1860.8	2121.8
Exports G&NFS	26.06	25.16	31.4	51.3	59.4	69.1	77.6	79.1	89.5	70.1	170.0	355.0	399.0	704.7	939.0	1566.7	1872.2
GDP	168.02	190.44	214.1	263.6	322.5	430.5	560.5	708.7	737.0	818.1	1070.4	1393.9	1833.8	2493.6	3319.1	4608.4	6474.4
Net Factor Income			-1.1	0.4	1.8	1.3	5.3	8.8	8.3	5.0	-2.4	-12.9	-9.3	-50.1	-55.9	-5.7	-72.3
Gross National Product			213.0	264.0	324.3	431.8	565.7	717.4	745.3	823.1	1068.0	1380.9	1824.5	2443.5	3263.3	4602.7	6402.2
Memo Items:																	
Resource Balance	-3.8	-10.4	-7.2	-7.9	-11.9	-23.7	-34.8	-48.3	-56.2	-57.3	-66.3	-69.7	-181.7	-191.4	-285.4	-294.1	-249.7
Expend. on Domestic Goods a/	142.0	165.3	182.7	212.3	263.1	361.4	482.9	629.6	647.6	748.1	900.5	1038.9	1434.8	1788.9	2380.2	3041.7	4602.3

a/ Expenditures minus imports.
Source: BCP and World Bank estimate
16-Oct-91

Table 2.6: PARAGUAY - NATIONAL ACCOUNTS BY EXPENDITURE, 1938-1990

(1982 Guaranies, Billions)

	1938	1939	1940	1941	1942	1943	1944	1945	1946	1947	1948	1949	1950	1951	1952	1953	1954	1955
Consumption	99.7	116.2	110.2	111.6	117.0	122.2	122.7	119.9	130.9	111.3	114.3	129.7	132.7	134.1	147.7	136.8	151.2	156.6
Private	92.1	108.9	103.8	105.2	109.5	112.6	113.8	110.6	123.6	103.8	103.9	121.3	123.1	124.4	137.8	125.7	139.8	146.1
Public	7.6	7.3	6.4	6.3	7.5	9.7	8.9	9.3	7.4	7.5	10.4	8.4	9.6	9.7	9.9	11.0	11.5	10.5
Investment	3.0	3.7	3.7	2.9	2.7	3.0	3.2	4.5	5.1	4.6	5.0	5.6	6.8	7.5	5.7	6.4	5.5	8.5
Fixed	3.0	3.7	3.7	2.9	2.7	3.0	3.2	4.5	5.1	4.6	5.0	5.6	5.3	5.8	5.0	5.9	6.4	7.5
Private																		
Public																		
Change in Inventories	0.0	0.0	0.0	0.0	0.0	0.0	0.0	0.0	0.0	0.0	0.0	0.0	1.5	1.8	0.7	0.5	-0.9	1.1
Total Expenditures	102.7	119.9	113.9	114.4	119.8	125.2	125.9	124.4	136.0	115.9	119.3	135.3	139.5	141.6	153.4	143.2	156.8	165.1
Imports G&NFS	14.6	17.2	14.3	17.3	16.1	20.0	17.1	29.9	31.4	14.8	21.2	18.5	22.7	31.2	50.5	29.9	37.2	31.7
Exports G&NFS	16.3	19.7	16.3	20.6	21.0	22.0	21.2	30.8	33.0	18.6	22.9	24.6	26.2	34.2	40.0	37.0	34.8	30.5
GDP	104.4	122.4	115.8	117.7	124.6	127.2	130.0	125.4	137.6	119.7	121.0	141.4	143.0	144.6	142.9	150.3	154.3	164.0
Terms of Trade Effect														4.4	4.6	8.8	-5.1	0.7	2.2
Gross Domestic Income b/													147.4	149.2	151.7	145.2	155.0	166.2
Net Factor Income																		
Gross National Product																		
Memo Items:																		
Resource Balance	1.7	2.5	2.0	3.3	4.9	2.0	4.1	0.9	1.6	3.8	1.8	6.1	3.5	3.0	-10.5	7.1	-2.5	-1.2
Expend. on Domestic Goods a/	88.1	102.7	99.5	97.1	103.7	105.2	108.8	94.6	104.7	101.1	98.1	116.8	116.8	110.4	102.9	113.3	119.6	133.5

a/ Expenditures minus imports.
b/ Exports & Imports have been deflated by the same price index since 1984. Therefore from this date on there is no difference between GDP & Gross Domestic Income.
Source: BCP and World Bank estimates.

Table 2.6: PARAGUAY - NATIONAL ACCOUNTS BY EXPENDITURE, 1938-1990
(1982 Guaranies, Billions)

	1956	1957	1958	1959	1960	1961	1962	1963	1964	1965	1966	1967	1968	1969	1970	1971	1972
Consumption	160.5	178.7	188.9	184.0	182.2	187.2	176.9	187.3	196.2	199.6	207.2	225.3	238.5	248.5	256.6	274.9	283.4
Private	149.2	165.0	173.2	167.9	166.5	171.2	162.0	171.9	180.1	182.8	187.4	204.7	215.7	223.9	230.9	248.9	256.7
Public	11.3	13.6	15.7	16.1	15.7	16.1	14.9	15.4	16.1	16.9	19.8	20.6	22.9	24.5	25.7	26.0	26.7
Investment	8.0	8.8	14.0	10.5	15.2	20.2	17.3	16.5	19.1	26.5	29.9	33.1	34.0	37.3	36.5	40.8	49.0
Fixed	6.8	8.0	13.4	11.0	14.5	18.1	15.2	14.4	17.0	24.6	29.0	32.4	32.7	35.6	35.9	39.2	44.2
Private																	
Public																	
Change in Inventories	1.2	0.7	0.6	-0.5	0.6	2.1	2.1	2.1	2.1	2.0	0.8	0.7	1.3	1.8	0.6	1.6	4.8
Total Expenditures	168.5	187.4	202.9	194.5	197.3	207.5	194.3	203.7	215.3	226.2	237.1	258.3	272.5	285.8	293.2	315.6	332.4
Imports G&NFS	29.2	40.5	38.2	30.8	32.1	31.7	19.3	18.2	22.2	27.1	30.3	31.1	35.6	42.0	39.9	44.4	44.7
Exports G&NFS	28.6	31.1	24.8	24.0	21.6	23.3	30.1	28.1	29.5	36.2	32.0	32.6	32.0	36.9	42.0	40.9	45.4
GDP	167.9	178.0	189.6	187.6	186.8	199.0	205.0	213.6	222.6	235.3	238.8	259.8	268.9	280.7	295.3	312.2	333.1
Terms of Trade Effect	0.9	3.6	5.0	3.9	4.0	2.7	-13.4	-11.8	-10.5	-10.5	-7.2	-8.2	-5.7	-5.6	-5.1	-3.7	-0.9
Gross Domestic Income b/	168.8	181.6	194.5	191.5	190.8	201.8	191.6	201.8	212.1	224.8	231.6	251.6	263.2	275.1	290.2	308.5	332.2
Net Factor Income																	
Gross National Product																	
Memo Items:																	
Resource Balance	-0.6	-9.4	-13.4	-6.9	-10.5	-8.4	10.7	9.9	7.3	9.1	1.7	1.5	-3.7	-5.1	2.2	-3.5	0.7
Expend. on Domestic Goods a/	139.3	146.9	164.7	163.7	165.2	175.8	174.9	185.5	193.1	199.1	206.8	227.2	236.9	243.8	253.3	271.3	287.7

a/ Expenditures minus imports.
b/ Exports & Imports have been deflacted by the same price index since 1984. Therefore from this date on there is no difference between GDP & Gross Domestic Income.
Source: BCP and World Bank estimates.

Table 2.6: PARAGUAY - NATIONAL ACCOUNTS BY EXPENDITURE, 1938-1990

(1982 Guaranies, Billions)

	1973	1974	1975	1976	1977	1978	1979	1980	1981	1982	1983	1984	1985	1986	1987	1988	1989	1990
Consumption	294.6	323.7	346.4	369.6	422.7	449.5	441.2	515.4	575.7	604.3	597.3	616.9	629.0	649.5	662.9	714.0	734.5	735.9
Private	269.2	300.1	317.4	339.4	388.2	409.3	403.7	472.6	524.6	552.0	543.4	566.2	576.2	595.6	603.5	652.7	669.8	669.2
Public	25.4	23.6	29.0	30.2	34.5	40.2	37.5	42.8	51.1	52.3	53.9	50.8	52.8	54.0	59.4	61.3	64.6	66.7
Investment	70.0	78.2	78.8	96.5	108.4	138.5	161.0	194.5	227.5	188.9	154.6	156.9	158.6	164.5	175.5	182.4	200.7	219.1
Fixed	58.8	66.9	63.5	87.4	103.8	126.2	150.8	183.7	216.8	176.9	145.2	146.4	146.9	151.5	161.0	167.0	184.8	203.4
Private					65.8	89.2	114.9	150.8	178.3	143.7	116.1	109.2	102.7	102.2	113.5	117.7	154.7	170.3
Public					38.1	37.0	35.9	32.9	38.5	33.2	29.1	37.3	44.2	49.4	47.5	49.2	30.1	33.1
Change in Inventories	11.1	11.3	15.3	9.2	4.5	12.4	10.2	10.8	10.7	12.0	9.4	10.5	11.7	13.0	14.5	15.4	16.0	15.7
Total Expenditures	364.5	401.8	425.2	466.1	531.1	588.0	602.2	709.8	803.2	793.2	751.9	773.9	787.7	814.0	838.4	896.4	935.2	955.0
Imports G&NFS	51.6	64.7	60.7	69.2	97.6	110.8	120.5	135.2	142.2	145.6	112.7	131.7	131.0	152.8	182.4	198.2	226.0	235.6
Exports G&NFS	45.1	51.3	52.0	49.7	62.0	74.5	132.7	110.1	83.4	89.5	75.7	94.7	109.5	105.0	143.5	152.0	190.2	207.8
GDP	358.0	388.5	416.5	446.7	495.5	551.7	614.4	684.7	744.4	737.0	714.9	736.9	766.2	766.2	799.4	850.2	899.5	927.3
Terms of Trade Effect	9.0	5.1	-9.0	6.5	22.6	17.8	-42.9	-16.7	4.9	0.0	-13.7	0.0	0.0	0.0	0.0	0.0	0.0	0.0
Gross Domestic Income b/	367.0	393.6	407.4	453.2	518.1	569.5	571.5	668.0	749.3	737.0	701.2	736.9	766.2	766.2	799.4	850.2	899.5	927.3
Net Factor Income					-1.3	0.5	8.4	10.1	8.4	8.3	8.5	-1.4	-4.0	-2.5	-10.2	-9.0	-0.7	-8.0
Gross National Product				494.2	552.3	622.8	694.8	752.7	745.3	723.4	735.6	762.2	763.7	789.2	841.2	898.8	919.3
Memo Items:																		
Resource Balance	-6.5	-13.4	-8.7	-19.5	-35.6	-36.3	12.2	-25.2	-58.8	-56.2	-37.0	-36.9	-21.5	-47.8	-39.0	-46.2	-35.7	-27.7
Expend. on Domestic Goods a/	312.9	337.2	364.4	396.9	433.5	477.3	481.7	574.6	661.0	647.6	639.2	642.2	656.7	661.3	655.9	698.2	709.3	719.5

a/ Expenditures minus imports.
b/ Exports & Imports have been deflacted by the same price index since 1984. Therefore from this date on there is no difference between GDP & Gross Domestic Income.
Source: BCP and World Bank estimates.

Table 2.7: PARAGUAY - NATIONAL ACCOUNTS BY EXPENDITURE, 1938-1990
(Price Indices, 1982=100)

	1938	1939	1940	1941	1942	1943	1944	1945	1946	1947	1948	1949	1950	1951	1952	1953	1954	1955
Consumption													1.0	1.7	3.0	5.9	7.1	8.9
Private													1.0	1.7	3.0	5.9	7.1	8.9
Public													1.1	1.7	3.6	5.9	7.5	9.3
Investment													1.6	2.5	5.7	9.1	13.0	13.2
Fixed													1.8	2.7	6.1	9.5	12.1	14.0
Private																		
Public																		
Change in Inventories													1.0	1.7	3.3	5.1	6.1	7.8
Total Expenditures													1.0	1.8	3.1	6.1	7.3	9.2
Imports G&NFS													1.1	1.7	2.5	6.1	7.3	8.8
Exports G&NFS													1.3	1.9	3.1	5.3	7.5	9.5
GDP													1.1	1.8	3.3	5.8	7.4	9.3
GDY b/													1.0	1.8	3.1	6.1	7.3	9.1
Net Factor Income																		
Gross National Product																		
Memo Items:																		
Expend. on Domestic Goods a/													1.0	1.8	3.4	6.0	7.4	9.2
Terms of Trade													116.7	113.4	121.9	86.2	102.0	107.3

a/ Expenditures minus imports.
b/ Exports & Imports have been deflacted by the same price index since 1984. Therefore from this date on there is no difference between GDP & Gross Domestic Income.
Source: BCP and World Bank estimates.

14-Jul-92

Table 2.7: PARAGUAY - NATIONAL ACCOUNTS BY EXPENDITURE, 1938-1990
(Price Indices, 1982 = 100)

	1956	1957	1958	1959	1960	1961	1962	1963	1964	1965	1966	1967	1968	1969	1970	1971	1972
Consumption	11.4	13.1	13.5	15.2	17.9	19.6	23.0	23.3	23.7	24.0	24.7	23.9	24.2	25.0	25.2	26.8	29.1
Private	11.4	13.1	13.4	15.2	18.0	19.7	23.1	23.4	23.8	24.1	24.8	24.0	24.3	24.9	25.1	26.8	29.1
Public	11.3	13.1	13.9	15.3	16.6	18.3	21.5	21.8	21.7	22.5	23.2	23.5	23.7	25.8	26.2	27.4	29.1
Investment	16.9	20.8	22.8	27.4	29.0	25.9	32.9	32.8	31.9	31.8	31.1	31.0	30.4	30.1	30.2	29.9	29.8
Fixed	18.0	21.6	23.3	26.6	29.5	26.1	34.5	34.4	33.2	32.5	31.3	31.2	30.7	30.4	30.3	30.1	30.0
Private																	
Public																	
Change in Inventories	10.4	12.0	12.6	10.4	19.1	23.6	21.5	21.8	21.7	22.5	23.2	23.5	23.7	24.2	24.0	25.1	27.5
Total Expenditures	11.6	13.5	14.1	15.8	18.7	20.2	23.9	24.1	24.4	24.9	25.5	24.8	25.0	25.7	25.9	27.2	29.2
Imports G&NFS	11.8	13.0	15.4	18.1	20.5	23.3	34.5	34.4	33.2	32.5	31.3	31.2	30.7	30.4	30.3	30.1	30.0
Exports G&NFS	12.1	14.5	18.5	21.0	24.3	26.0	19.1	19.9	21.4	23.1	24.3	23.3	25.2	25.8	26.6	27.4	29.4
GDP	11.7	13.8	14.4	16.1	19.0	20.4	22.2	22.6	23.1	23.8	24.6	23.9	24.3	25.0	25.4	26.8	29.1
GDY b/	11.6	13.5	14.1	15.8	18.7	20.1	23.7	24.0	24.3	24.9	25.3	24.7	24.8	25.5	25.8	27.1	29.2
Net Factor Income																	
Gross National Product																	
Memo Items:																	
Expend. on Domestic Goods a/	11.6	13.6	13.8	15.4	18.4	19.6	22.7	23.1	23.4	23.9	24.6	24.0	24.1	24.8	25.2	26.7	29.0
Terms of Trade	103.1	111.6	119.9	116.2	118.4	111.6	55.4	57.8	64.5	71.1	77.6	74.7	82.1	84.9	87.8	91.0	98.0

a/ Expenditures minus imports.
b/ Exports & Imports have been deflected by the same price index since 1984. Therefore from this date on there is no difference between GDP & Gross Domestic Income.
Source: BCP and World Bank estimates.

14-Jul-92

Table 2.7: PARAGUAY - NATIONAL ACCOUNTS BY EXPENDITURE, 1938-1990
(Price Indices, 1982 = 100)

	1973	1974	1975	1976	1977	1978	1979	1980	1981	1982	1983	1984	1985	1986	1987	1988	1989	1990
Consumption	34.2	42.2	44.7	45.6	48.8	54.9	75.1	84.2	96.0	100.0	117.2	144.5	184.0	239.7	310.7	391.6	517.9	712.5
Private	34.4	42.4	45.0	45.7	49.0	55.0	75.9	84.5	96.1	100.0	118.2	145.2	185.2	240.9	311.9	396.3	522.1	723.5
Public	32.2	39.1	41.3	44.5	47.4	53.4	65.9	81.1	95.1	100.0	107.6	136.5	170.9	225.7	297.6	341.1	474.5	602.6
Investment	34.1	45.1	58.2	54.6	60.0	63.3	76.4	82.9	89.8	100.0	113.4	156.4	193.2	279.0	356.6	443.4	547.4	675.7
Fixed	34.7	46.2	62.3	55.8	60.6	64.4	77.0	83.1	89.6	100.0	113.3	157.9	196.0	285.0	367.4	460.1	565.9	700.8
Private					61.0	64.5	77.5	83.1	89.7	100.0	112.7	158.4	196.8	300.1	365.6	467.1	569.2	704.9
Public					59.8	64.1	75.3	83.2	89.1	100.0	115.7	156.4	193.9	253.9	371.6	443.4	548.9	679.7
Change in Inventories	31.0	38.8	41.4	43.3	47.3	52.3	67.1	79.4	93.7	100.0	114.0	136.2	158.1	208.2	237.2	262.9	333.2	350.1
Total Expenditures	34.2	42.8	47.2	47.5	51.1	56.9	75.4	83.9	94.2	100.0	116.4	146.9	185.8	247.6	320.3	402.1	524.2	704.1
Imports G&NFS	34.7	46.2	58.5	55.8	60.6	64.4	77.0	83.1	89.6	100.0	113.0	179.4	324.3	380.1	491.2	617.6	823.5	900.8
Exports G&NFS	41.6	50.8	48.4	63.1	82.7	79.8	52.1	70.5	94.9	100.0	92.5	179.4	324.3	380.1	491.2	617.6	823.5	900.8
GDP	35.0	43.3	45.7	47.9	53.2	58.5	70.1	81.9	95.2	100.0	114.4	145.3	181.9	239.3	311.9	390.4	512.3	698.2
GDY b/	34.2	42.7	46.7	47.2	50.9	56.6	75.3	83.9	94.6	100.0	116.7	145.3	181.9	239.3	311.9	390.4	512.3	698.2
Net Factor Income					-27.8	340.3	15.3	51.8	104.4	100.0	58.7	179.4	324.3	368.4	491.2	617.6	823.1	901.2
Gross National Product					53.4	58.7	69.3	81.4	95.3	100.0	113.8	145.2	181.2	238.9	309.6	387.9	512.1	696.4
Memo Items:																		
Expend. on Domestic Goods a/	34.1	42.1	45.4	46.0	49.0	55.1	75.0	84.0	95.2	100.0	117.0	140.2	158.2	217.0	272.7	340.9	428.9	639.7
Terms of Trade	119.9	110.0	82.6	113.1	136.5	123.9	67.7	84.8	105.9	100.0	81.9	100.0	100.0	100.0	100.0	100.0	100.0	100.0

a/ Expeditures minus imports
b/ Exports & Imports have been deflacted by the same price index since 1984. Therefore from this date on there is no difference between GDP & Gross Domestic Income.

Source: BCP and World Bank estimates.

Table 2.8: PARAGUAY - NATIONAL ACCOUNTS BY EXPENDITURE, 1938-1990
(Growth Rates, Percent)

	1938	1939	1940	1941	1942	1943	1944	1945	1946	1947	1948	1949	1950	1951	1952	1953	1954	1955
Consumption		16.5	-5.2	1.2	4.9	4.5	0.4	-2.3	9.2	-15.0	2.7	13.5	2.3	1.0	10.1	-7.4	10.6	3.5
Private		18.3	-4.7	1.4	4.1	2.8	1.1	-2.8	11.7	-16.0	0.1	16.8	1.5	1.0	10.8	-8.7	11.2	4.5
Public		-4.5	-12.6	-0.8	18.3	29.2	-8.0	4.3	-20.8	2.4	38.5	-19.8	14.7	0.9	2.3	11.4	3.9	-8.7
Investment		22.9	0.2	-21.6	-4.4	7.8	8.6	40.8	12.8	-10.3	8.7	12.8	20.4	11.4	-24.6	13.0	-13.7	54.1
Fixed		22.9	0.2	-21.6	-4.4	7.8	8.6	40.8	12.8	-10.3	8.7	12.8	-6.3	9.5	-14.1	18.6	9.4	16.0
Private																		
Public																		
Change in Inventories														18.1	-58.9	-24.7	-260.6	223.5
Total Expenditures		16.7	-5.0	0.5	4.7	4.5	0.6	-1.2	9.3	-14.8	2.9	13.5	3.1	1.5	8.3	-6.6	9.5	5.3
Imports G&NFS		17.5	-16.6	20.7	-7.1	24.3	-14.4	74.7	5.0	-52.8	43.1	-12.6	22.6	37.7	61.7	-40.8	24.5	-15.0
Exports G&NFS		20.9	-17.2	26.1	1.9	5.1	-4.0	45.6	7.0	-43.6	23.3	7.2	6.6	30.5	17.1	-7.5	-6.1	-12.3
GDP		17.2	-5.3	1.6	5.9	2.1	2.1	-3.5	9.8	-13.0	1.1	16.8	1.1	1.1	-1.2	5.2	2.7	6.2
Gross Domestic Income b/														1.2	1.7	-4.3	6.8	7.2
Net Factor Income																		
Gross National Product																		
Memo Items:																		
Resource Balance		50.2	-21.6	66.1	49.8	-58.2	98.6	-77.3	73.3	138.8	-53.7	245.2	-42.0	-15.8	-451.9	168.0	-134.8	53.0
Expend. on Domestic Goods a/		16.6	-3.1	-2.4	6.7	1.5	3.4	-13.1	10.7	-3.4	-3.0	19.1	0.0	-5.5	-6.8	10.1	5.5	11.6

a/ Expenditures minus imports.
b/ Exports & Imports have been deflacted by the same price index since 1984. Therefore from this date on there is no difference between GDP & Gross Domestic Income.
Source: BCP and World Bank estimates.

Table 2.8: PARAGUAY - NATIONAL ACCOUNTS BY EXPENDITURE, 1938-1990
(Growth Rates, Percent)

	1956	1957	1958	1959	1960	1961	1962	1963	1964	1965	1966	1967	1968	1969	1970	1971	1972
Consumption	2.5	11.3	5.7	-2.6	-1.0	2.8	-5.5	5.8	4.8	1.7	3.8	8.7	5.9	4.2	3.3	7.1	3.1
Private	2.1	10.6	5.0	-3.1	-0.8	2.8	-5.3	6.1	4.8	1.5	2.5	9.2	5.4	3.8	3.1	7.8	3.1
Public	8.0	20.6	14.9	2.8	-2.5	2.1	-7.2	3.4	4.8	4.4	17.4	3.9	11.3	7.2	4.8	1.0	2.8
Investment	-6.4	9.6	60.3	-25.3	44.3	33.6	-14.4	-4.9	16.0	38.9	12.5	10.8	2.8	9.8	-2.1	11.6	20.2
Fixed	-8.6	17.9	66.8	-17.9	32.0	24.6	-16.2	-5.4	18.4	44.6	18.1	11.5	1.0	8.8	1.0	9.1	12.8
Private																	
Public																	
Change in Inventories	8.4	-38.4	-12.2	-179.8	222.0	243.1	0.5	-1.2	-0.6	-6.7	-57.2	-13.4	79.7	34.9	-64.3	150.2	205.0
Total Expenditures	2.1	11.2	8.3	-4.1	1.5	5.1	-6.4	4.9	5.7	5.0	4.8	9.0	5.5	4.9	2.6	7.7	5.3
Imports G&NFS	-7.7	38.5	-5.6	-19.3	4.2	-1.3	-39.1	-5.6	21.8	21.9	11.8	2.7	14.6	17.8	-5.0	11.3	0.8
Exports G&NFS	-6.3	8.7	-20.1	-3.5	-9.9	7.9	29.0	-6.5	4.9	22.8	-11.6	1.7	-1.8	15.3	13.9	-2.7	11.0
GDP	2.4	6.0	6.5	-1.0	-0.4	6.6	3.0	4.2	4.2	5.7	1.5	8.8	3.5	4.4	5.2	5.7	6.7
Gross Domestic Income b/	1.6	7.6	7.1	-1.5	-0.4	5.8	-5.0	5.3	5.1	6.0	3.0	8.6	4.6	4.6	5.5	6.3	7.7
Net Factor Income																	
Gross National Product																	
Memo Items:																	
Resource Balance	44.7	-1363.2	-42.2	48.6	-53.6	20.0	227.0	-8.0	-26.4	25.5	-81.0	-15.3	-350.0	-39.1	142.4	-261.4	119.4
Expend. on Domestic Goods a/	4.4	5.5	12.1	-0.6	0.9	6.4	-0.5	6.0	4.1	3.1	3.9	9.9	4.3	2.9	3.9	7.1	6.1

a/ Expenditures minus imports.
b/ Exports & Imports have been deflacted by the same price index since 1984. Therefore from this date on there is no difference between GDP & Gross Domestic Income.
Source: BCP and World Bank estimates.

Table 2.8: PARAGUAY - NATIONAL ACCOUNTS BY EXPENDITURE, 1938-1990
(Growth Rates, Percent)

	1973	1974	1975	1976	1977	1978	1979	1980	1981	1982	1983	1984	1985	1986	1987	1988	1989	1990
Consumption	4.0	9.9	7.0	6.7	14.4	6.3	-1.8	16.8	11.7	5.0	-1.1	3.3	2.0	3.3	2.1	7.7	2.9	0.2
Private	4.9	11.5	5.8	6.9	14.4	5.4	-1.4	17.1	11.0	5.2	-1.6	4.2	1.8	3.4	1.3	8.2	2.6	-0.1
Public	-5.0	-7.0	22.6	4.2	14.5	16.6	-6.8	14.2	19.4	2.2	3.2	-5.9	4.0	2.2	10.0	3.3	6.4	3.2
Investment	42.7	11.7	0.8	22.5	12.3	27.8	16.2	20.8	17.0	-17.0	-18.2	1.5	1.1	3.7	6.7	3.9	10.1	9.1
Fixed	33.0	13.7	-5.1	37.6	18.9	21.5	19.5	21.8	18.0	-18.4	-17.9	0.9	0.3	3.1	6.3	3.7	10.7	10.1
Private						35.6	28.8	31.2	18.2	-19.4	-19.2	-6.0	-5.9	-0.5	11.1	3.7	31.4	10.1
Public						-2.9	-2.8	-8.4	16.9	-13.8	-12.3	28.2	18.7	11.5	-3.7	3.7	-38.9	10.1
Change in Inventories	131.9	1.3	36.1	-40.2	-50.4	171.8	-17.6	5.8	-0.2	12.1	-21.9	11.7	11.4	11.1	11.5	6.5	3.5	-1.7
Total Expenditures	9.7	10.2	5.8	9.6	13.9	10.7	2.4	17.9	13.2	-1.2	-5.2	2.9	1.8	3.3	3.0	6.9	4.3	2.1
Imports G&NFS	15.4	25.3	-6.1	14.0	41.1	13.5	8.8	12.2	5.1	2.4	-22.6	16.8	-0.5	16.6	19.4	8.7	14.0	4.3
Exports G&NFS	-0.6	13.7	1.4	-4.4	24.7	20.1	78.2	-17.0	-24.3	7.3	-15.3	25.1	15.6	-4.1	36.7	6.0	25.1	9.3
GDP	7.5	8.5	7.2	7.3	10.9	11.4	11.4	11.4	8.7	-1.0	-3.0	3.1	4.0	0.0	4.3	6.4	5.8	3.1
Gross Domestic Income b/	10.5	7.2	3.5	11.2	14.3	9.9	0.3	16.9	12.2	-1.6	-4.9	5.1	4.0	0.0	4.3	6.4	5.8	3.1
Net Factor Income						-140.3	1520.3	20.2	-17.4	-1.5	2.9	-115.9	195.3	-37.0	305.1	-11.3	-92.3	1055.6
Gross National Product						11.7	12.8	11.6	8.3	-1.0	-2.9	1.7	3.6	0.2	3.3	6.6	6.9	2.3
Memo Items:																		
Resource Balance	-1059.2	-106.6	34.7	-123.1	-82.9	-2.0	133.5	-306.7	-133.9	4.5	34.1	0.2	41.8	-122.4	18.5	-18.6	22.7	22.4
Expend. on Domestic Goods a/	8.8	7.8	8.1	8.9	9.2	10.1	0.9	19.3	15.0	-2.0	-1.3	0.5	2.3	0.7	-0.8	6.4	1.6	1.4

a/ Expenditures minus imports.
b/ Exports & Imports have been deflacted by the same price index since 1984. Therefore from this date on there is no difference between GDP & Gross Domestic Income.
Source: BCP and World Bank estimates.

TABLE 2.9: PARAGUAY - Harvested Area, Production Volume, and Yield of Selected Agricultural Products, 1971-1990

	1971	1972	1973	1974	1975	1976	1977	1978	1979	1980	1981	1982	1983	1984	1985	1986	1987	1988	1989	1990
Area, thousands of hectares																				
Grains	375.6	374.7	363.0	453.9	498.8	563.1	679.1	712.4	894.2	1022.6	820.0	1142.1	1134.3	1283.4	1432.7	1168.2	1630.0	1643.0	1668.7	1745.6
Soybeans	54.6	76.8	81.4	127.3	150.2	173.4	228.8	272.2	360.3	476.3	398.1	502.2	567.8	638.8	718.8	539.9	974.0	766.0	861.0	900.0
Maize	190.1	184.4	195.6	206.1	222.6	257.3	282.1	275.8	352.7	375.6	290.8	369.2	399.1	435.6	470.4	376.2	567.0	486.0	500.0	518.0
Others a/	130.9	114.5	96.0	120.5	126.0	132.4	168.2	164.3	181.2	170.7	133.1	270.7	167.4	209.0	243.5	252.1	289.0	291.0	337.7	327.0
Other crops	235.1	263.4	260.0	293.7	304.8	336.0	438.6	527.2	560.5	517.9	528.0	536.2	560.5	602.0	701.1	731.3	692.0	783.6	767.0	881.0
Cotton	33.2	57.2	81.1	93.2	100.0	108.9	200.2	284.9	312.5	268.3	244.0	248.1	262.8	294.0	395.9	384.8	284.4	403.0	439.0	509.0
Sugar	39.7	28.9	28.0	28.7	30.3	31.0	33.1	34.9	34.8	36.6	48.6	48.9	51.8	54.6	65.2	59.1	63.5	63.0	57.5	47.0
Maniac	94.6	93.3	79.5	90.1	96.5	106.6	113.5	120.3	128.4	135.7	179.2	179.5	180.7	183.5	166.4	199.6	205.2	229.5	234.1	240.0
Others b/	67.6	74.0	71.4	81.7	78.0	89.6	91.8	97.1	86.8	97.3	57.2	60.7	65.4	69.9	73.6	88.0	106.9	98.1	37.4	85.0
Total	610.7	628.1	623.0	747.6	803.6	899.1	1117.7	1239.6	1454.7	1540.5	1348.0	1678.3	1694.8	1885.4	2133.8	1899.5	2182.0	2326.6	2455.7	2626.0
Volume, thousands of tons																				
Grains	432.9	409.6	476.9	602.1	656.1	785.3	963.8	866.4	1289.3	1317.5	1394	1508.2	1694.7	1982.1	2316.9	1505.5	2672.4	2863.6	3317.8	3542.6
Soybeans	76.3	97.1	122.6	181.3	220.1	283.5	376.9	333.1	549.2	673.9	761.2	755.5	849.7	975.4	1172.5	692.3	1178.6	1407.4	1614.6	1794.9
Maize	230.5	209.3	246.1	261.6	300.8	351.5	401	365.4	550.4	584.7	485.2	552.8	619.5	730.2	800.8	499.5	1001.4	960.59	1000.4	1138.8
Others a/	127.1	103.2	108.2	139.2	135.2	150.3	185.9	177.9	189.7	158.9	174.6	199.1	225.5	276.5	343.6	374.7	492.4	495.61	702.8	609.1
Other crops	2835	2460	2460.2	2905.4	2654.5	2962.9	3392	3803.6	3820.7	3929.4	4642.2	5245.2	5422.6	5829.2	6245.1	6465.1	7329.6	7491.2	7984.9	8754.7
Cotton	17.5	52.9	85.2	89.7	99.8	107.5	227.4	283.4	234.7	227.5	341.8	260.4	235.8	319.9	469.3	343.2	248.3	543.2	630.2	642.6
Sugar	1407.4	1045	1100.8	1203	1038.2	1076.8	1159.7	1260.1	1287	1372.6	2154.7	2319.4	2406.7	2550.2	2726.5	2768	3187.7	2668.1	2868.7	2256.1
Maniac	1195.8	1208	1108.3	1395.1	1427.6	1673.3	1760.3	1837.5	1888	2031	2012.4	2511.2	2610	2776.2	2861.3	2876.3	3467.2	3890.9	3978.3	3649.95
Others b/	214.3	154.1	165.9	217.6	89.1	225.2	244.6	222.6	211	298	133.6	154.2	170.1	183.9	188	466.6	426.3	389	407.7	306.05
Total	3267.9	2869	2937.1	3507.5	3310.6	3768.2	4355.8	4470	4910	5246.9	6036.2	6753.4	7117.3	7811.3	8562	7970.6	10002	10354.8	11202.7	10297.3
Livestock (thous. slaughters)	695.2	686.9	571.0	577.9	498.3	537.3	635.3	596.2	577.8	565.1	533.9	644.0	567.8	530.8	539.3	550.1	543.8	577.9	612.0	
Yield (ton/ha)																				
Grains	1.15	1.09	1.31	1.33	1.32	1.39	1.42	1.22	1.44	1.29	1.70	1.32	1.49	1.54	1.62	1.29	1.75	1.66	1.98	2.03
Soybeans	1.38	1.26	1.51	1.42	1.47	1.63	1.65	1.22	1.52	1.21	1.90	1.51	1.50	1.53	1.63	1.23	1.76	1.84	1.90	1.99
Maize	1.21	1.14	1.33	1.37	1.35	1.37	1.42	1.29	1.56	1.55	1.61	1.50	1.55	1.66	1.70	1.25	1.77	1.98	2.00	2.20
Others a/	0.97	0.90	1.13	1.18	1.07	1.14	1.11	1.08	1.05	0.93	1.31	0.74	1.35	1.32	1.41	1.48	1.70	1.70	2.08	1.86
Other crops	12.06	9.71	9.46	9.89	8.71	8.88	7.73	6.84	6.46	7.59	8.79	9.78	9.67	9.88	8.91	8.84	11.07	9.56	10.28	7.87
Cotton	0.53	0.92	1.05	0.96	1.00	0.98	1.14	0.99	0.75	0.89	1.40	1.06	0.90	1.09	1.22	0.89	0.87	1.35	1.44	1.26
Sugar	35.45	36.14	39.31	41.92	34.26	34.74	35.04	36.11	36.98	37.51	44.34	48.48	48.48	46.71	49.39	46.67	50.20	50.34	49.89	48.00
Maniac	12.64	12.96	13.94	15.48	14.79	14.77	15.51	15.27	14.94	14.97	11.29	13.99	14.44	15.12	15.36	14.41	16.90	16.95	16.99	14.79
Others b/	3.17	2.08	2.32	2.66	1.14	2.54	2.66	2.56	2.43	3.41	2.33	2.54	2.60	2.63	2.55	5.55	3.81	3.97	10.90	3.60
Total	5.35	4.57	4.71	4.69	4.12	4.19	3.90	3.61	3.38	3.41	4.48	4.02	4.20	4.14	4.01	4.20	4.58	4.45	4.56	3.92

Source: Government of Paraguay, Ministry of Agriculture.

a/ Includes rice, peas, beans, wheat.
b/ Includes sweet potato, potato, onion, peanuts, tobacco.

Table 2.10: Paraguay - Domestic Prices for Selected Agricultural Goods, 1970-1989
(Producer Prices, Guaranies per kilo, Year Average)

	1970	1971	1972	1973	1974	1975	1976	1977	1978	1979	1980	1981	1982	1983	1984	1985	1986	1987	1988	1989
Soybeans	7.5	8.5	11	22	20	18	20	21	21	24	19	28	25	30	61	51	86	78	167	187
Cotton																				
First Grade	13.5	17	21	22	36	27	44	46	43	49	59	54	47	87	123	115	145	258	274	397
Second Grade	11	14	17	19	30	25	37	43	42	44	56	50	40	--	--	100	133	210	243	393
Corn																				
White	5	5	8	12	15	13	15	16	20	24	25	22	22	84	65	63	256	151	116	151
V-1	4	4	6.5	9	11	10	11	10	11	17	16	15	14	33	30	39	61	56	86	78
Sugar	616	630	646	823	1110	1354	1730	1750	1810	1948	2500	2120	2650	2650	3240	3930	5400	8000	10000	16000
Manioc	3	3	3.5	4	6	7	9	10	9	10	13	13	10	10	14	16	34	28	45	47
Coffee	60	60	53	94	105	127	150	210	393	310	400	300	120	167	--	--	--	--	--	744
Wheat	9.5	10	9.7	12	20	25	25	22	22	24	25	26	35	45	51	63	74	80	95	122
Yerba Mate	6	6.5	6.5	10	19	17	17	35	30	40	54	70	40	71	95	130	265	346	546	522
Memo:																				
Beef (live)	19	22	31	43	48	41	35	44	47	69	96	100	90	90	130	177	264	352	350	343
Aromatic Oils																				
Mint	--	--	628	550	900	1320	686	1200	1100	850	--	--	3000	1450	1900	2350	--	3040	7650	23500
Petit-grain	366	428	571	1318	1700	620	600	720	800	932	1100	1850	1630	1300	1500	1950	2004	3417	4633	10502
Planting Seeds																				
Cotton	8	8.5	13	18	27	27	28	35	34	37	42	42	44	50	70	93	99	164	268	426
Soybeans	11	13	16	36	76	33	38	42	34	47	40	65	62	150	158	178	216	270	250	298
Corn	7	7	14	19	17	17	20	34	33	32	38	46	44	50	112	191	401	363	270	332
Wheat	14	15.5	15	17	29	38	42	47	35	51	50	55	63	60	75	106	150	270	150	180

Source: Government of Paraguay, Ministry of Agriculture

TABLE 2.11: PARAGUAY - INDUSTRIAL SECTOR, VALUE ADDED, by subsectors
(Current Guaranies, Millions)

	1976	1977	1978	1979	1980	1981	1982	1983	1984	1985	1986	1987	1988	1989
Food, Beverage, Tobacco	14886	18405	21368	24867	31780	40520	48023	56380	76307	106369	152186	206246	260648	312651
Clothing, Textiles, Leather, Footwear	5006	8513	10453	10694	10349	15194	12473	13036	22086	28903	32408	40245	91237	105415
Wood Products	2643	3480	5103	9380	13916	5522	16335	17421	20646	23305	31301	38193	46378	87429
Furniture	351	392	421	750	916	1212	1343	1855	2302	2234	3340	4339	5728	9879
Paper	35	39	43	55	69	92	105	110	132	168	249	321	397	468
Printing	672	945	1019	2324	3437	4423	6229	6299	5945	7047	9982	11946	17702	25181
Chemicals	1410	1464	1645	1819	2152	2240	2315	2628	4318	5782	5693	9293	9225	17267
Petroleum Refining	3634	4899	5293	8450	13692	17558	9508	9190	8686	11966	13616	22320	33198	76169
Rubber and Plastics	213	255	273	437	865	1635	2043	2424	3045	3169	3816	4918	5881	14293
Ceramics and Glassware	17	18	19	49	83	203	272	464	531	1057	672	1428	1906	4015
Nonmetal Minerals	1490	1890	2045	3336	3743	4753	4254	5310	5239	5368	11557	16222	19265	32598
Iron and Steel	37	13	17	14	19	27	33	41	53	67	91	963	2038	2575
Nonferrous Metals	29	30	35	76	153	321	413	489	610	678	905	1132	1562	3370
Metal Products	684	980	1159	1157	1310	1918	2116	2049	2979	3600	4574	5726	7647	13898
Machinery	80	90	93	140	179	246	314	371	459	589	789	1004	1427	1990
Transportation Equipment	323	375	406	565	732	1027	1263	1528	2906	3357	4399	5267	6450	13052
Other Manufactures	176	199	222	350	466	703	967	1181	1522	1684	1853	2311	2781	6128
Subtotal, Manufactures	31686	41987	49614	64463	83861	107594	108006	120773	157766	205343	277431	371874	513470	726378
Artesanry	2535	2987	4805	5147	8477	10875	12960	13500	14237	20772	18577	32677	42618	59211
TOTAL	34221	44974	54419	69610	92338	118469	120966	134273	172003	226115	296008	404551	556088	785589

Source: CBP

TABLE 2.12: PARAGUAY - INDUSTRIAL SECTOR, VALUE ADDED, by subsectors
(Constant 1982 Guaranies, Millions)

	1976	1977	1978	1979	1980	1981	1982	1983	1984	1985	1986	1987	1988	1989
Food, Beverage, Tobacco	30557	34341	36067	37794	40025	45456	48023	49523	52664	53588	57195	59359	51490	54429
Clothing, Textiles, Leather, Footwear	8873	11548	12534	11098	11276	13540	12473	11624	12523	15352	12783	10300	17725	18127
Wood Products	6116	7392	8917	14994	21861	16382	16335	15347	15034	13578	13899	15573	16369	17229
Furniture	831	853	890	950	1118	1273	1343	1630	1690	1724	1762	1831	1723	1686
Paper	62	64	63	69	80	97	105	96	99	103	118	127	124	125
Printing	3484	3899	4197	4985	5181	5788	6229	3777	4308	4384	4018	4090	4386	4422
Chemicals	2053	1881	1939	1919	2129	2008	2315	2410	3010	3521	2812	3107	2988	3050
Petroleum Refining	9377	12607	15222	15645	16936	16622	9508	9134	7084	8068	7912	9948	12773	13227
Rubber and Plastics	1061	1142	1285	1628	1751	1976	2043	1997	2376	2417	2114	2229	2567	2446
Ceramics and Glassware	62	62	39	147	214	247	272	460	485	724	567	587	678	686
Nonmetal Minerals	3176	3707	3517	3646	4431	4793	4254	4414	4070	3279	4694	5489	5629	5627
Iron and Steel	17	18	21	26	28	32	33	40	41	54	50	308	356	462
Nonferrous Metals	213	228	256	322	349	398	413	405	478	496	503	518	544	539
Metal Products	1330	1612	1616	1870	2125	2192	2116	2070	2394	2717	2432	2504	2390	2461
Machinery	175	187	208	252	270	301	314	310	363	462	432	442	453	467
Transportation Equipment	657	707	796	1010	1086	1225	1263	1257	1870	2853	2472	2287	2132	2251
Other Manufactures	504	543	610	774	832	939	967	970	1186	1364	1030	1059	992	1026
Subtotal, Manufactures	68548	80791	88177	97129	109692	113269	108006	105464	109675	114684	114793	119758	123319	128260
Artesanry	5270	5835	8535	9160	10730	12344	12960	10397	11400	12445	10552	9974	13990	17150
TOTAL	73818	86626	96712	106289	120422	125613	120966	115861	121075	127129	125345	129732	137309	145410

Source: CBP

TABLE 2.13: PARAGUAY - GROSS DOMESTIC INVESTMENT, 1962-1989
(Millions)

	1962	1963	1964	1965	1966	1967	1968	1969	1970	1971	1972	1973	1974	1975	1976
Current Guaranies															
Gross Dom. Fixed Investment	5,237	4,939	5,643	7,988	9,085	10,094	10,036	10,813	10,883	11,800	13,270	20,411	30,897	39,543	48,746
Buildings	2,669	2,945	3,167	3,573	4,325	4,909	4,503	5,100	5,463	6,380	6,670	9,010	14,061	18,850	23,922
Transportation	1,006	774	908	1,234	1,423	1,406	1,846	2,538	1,806	1,795	1,890	2,809	5,320	8,178	9,615
Communications	182	123	228	338	643	567	523	234	379	372	970	1,369	1,620	2,039	2,685
Machinery	1,381	1,097	1,340	2,843	2,694	3,212	3,164	2,941	3,235	3,253	3,740	7,223	9,896	10,476	12,524
Change in stock	460	462	457	442	195	171	310	427	151	395	1,320	3,451	4,374	6,350	3,970
Private Investment	4,731	4,335	5,106	7,245	6,807	6,438	5,762	7,299	8,079	8,962	10,350	18,831	29,912	36,228	33,738
Public Investment	967	1,066	994	1,185	2,473	3,827	4,584	3,941	2,955	3,233	4,240	5,031	5,359	9,665	18,978
Gross Domestic Investment	5,698	5,401	6,100	8,430	9,280	10,265	10,346	11,240	11,034	12,195	14,590	23,862	35,271	45,893	52,716
Constant 1982 Guaranies															
Gross Dom. Fixed Investment	15,180	14,358	17,000	24,578	29,026	32,353	32,691	35,569	35,917	39,203	44,233	58,821	66,877	63,472	87,358
Buildings	7,742	8,615	9,520	11,060	13,642	15,853	14,711	16,717	17,959	21,170	22,117	25,881	30,763	30,466	42,871
Transportation	2,884	2,154	2,720	3,687	4,644	4,529	5,884	8,537	6,106	5,880	6,193	8,235	11,369	13,329	17,231
Communications	455	431	680	983	2,032	1,618	1,635	711	1,077	1,176	3,538	4,117	3,344	3,174	4,812
Machinery	4,099	3,158	4,080	8,848	8,708	10,353	10,461	9,604	10,775	10,977	12,385	20,588	21,401	16,503	22,444
Change in stock	2,203	2,169	2,102	1,965	840	728	1,307	1,763	628	1,575	4,800	11,132	11,273	15,338	9,169
Private Investment	14,428	13,255	15,988	22,800	21,892	20,742	18,937	24,228	26,751	29,970	34,813	55,263	66,271	61,155	61,699
Public Investment	2,955	3,272	3,114	3,743	7,974	12,339	15,061	13,104	9,794	10,806	14,220	14,690	11,879	17,655	34,828
Gross Domestic Investment	17,383	16,527	19,102	26,543	29,866	33,081	33,998	37,332	36,545	40,778	49,033	69,953	78,150	78,810	96,527

Source: Central Bank of Paraguay

TABLE 2.13: PARAGUAY - GROSS DOMESTIC INVESTMENT, 1962-1989
(Millions)

	1977	1978	1979	1980	1981	1982	1983	1984	1985	1986	1987	1988	1989
Current Guaranies													
Gross Dom. Fixed Investment	62,922	81,256	116,142	152,654	194,219	176,871	164,509	231,178	287,951	431,810	591,415	768,163	1,045,592
Buildings	27,800	40,710	61,066	90,308	123,000	115,632	128,351	133,746	165,776	220,274	292,084	326,746	518,513
Transportation	16,580	22,556	25,669	30,489	29,467	19,278	6,855	59,891	58,705	99,610	155,162	274,132	287,839
Communications	2,072	2,128	1,104	2,541	5,092	6,639	8,175	2,922	9,586	61,987	84,888	118,267	139,969
Machinery	16,470	15,862	28,303	29,316	36,660	35,322	21,128	34,619	53,884	49,939	59,281	49,018	99,271
Change in stock	2,150	6,461	6,830	8,550	10,064	12,045	10,720	14,300	18,500	27,072	34,400	40,576	53,240
Private Investment	42,297	64,034	95,918	133,804	169,983	155,735	141,585	187,159	220,645	333,577	449,290	590,379	933,555
Public Investment	22,775	23,683	27,054	27,400	34,300	33,181	33,644	58,319	85,806	125,305	176,525	218,360	165,277
Gross Domestic Investment	65,072	87,717	122,972	161,204	204,283	188,916	175,229	245,478	306,451	458,882	625,815	808,739	1,098,832
Constant 1982 Guaranies													
Gross Dom. Fixed Investment	103,832	126,174	150,834	183,699	216,762	176,871	145,183	146,440	146,940	151,500	160,990	166,951	184,771
Buildings	45,875	63,215	79,307	108,673	137,277	115,632	113,185	91,208	90,296	91,200	93,024	95,484	92,916
Transportation	27,360	35,025	33,336	36,690	32,887	19,278	6,066	33,389	27,208	34,845	24,307	19,148	21,599
Communications	3,419	3,304	1,434	3,058	5,683	6,639	7,235	1,628	4,462	7,275	12,069	7,936	16,996
Machinery	27,178	24,630	36,757	35,278	49,915	35,322	18,967	20,215	24,974	18,180	31,590	44,383	53,260
Change in stock	4,545	12,354	10,179	10,766	10,741	12,045	9,404	10,500	11,700	13,000	14,500	15,436	15,976
Private Investment	70,317	101,571	125,080	161,534	189,013	155,735	125,502	119,655	114,393	115,150	127,986	133,143	170,635
Public Investment	38,060	36,957	35,933	32,931	38,490	33,181	29,085	37,285	44,247	49,350	47,504	49,244	30,112
Gross Domestic Investment	108,377	138,528	161,013	194,465	227,503	188,916	154,587	156,940	158,640	164,500	175,490	182,387	200,747

Source: Central Bank of Paraguay

TABLE 3.1: PARAGUAY - BALANCE OF PAYMENTS, 1980-1990
(US$ Millions)

	1980	1981	1982	1983	1984	1985	1986	1987	1988 1/	1988	1989	1990
Exports	944.9	965.6	781.6	742.2	756.2	615.5	759.2	794.0	1077.7	1017.9	1317.4	1579.0
Goods	757.5	757.5	619.2	554.8	560.4	466.0	575.0	595.0	871.0	871.0	1166.5	1392.3
Non-factor Services 2/	187.4	208.1	162.4	187.4	195.8	149.5	184.2	199.0	206.7	146.9	150.9	186.7
Imports	1193.5	1175.5	919.8	667.5	907.3	865.6	1139.0	1249.0	1373.8	1320.8	1297.6	1739.0
Goods	1048.0	1010.0	757.0	539.0	700.0	659.0	879.0	933.0	1063.0	1030.1	1001.3	1353.6
Non-factor Services 2/	145.5	165.5	162.8	128.5	207.3	206.6	260.0	316.0	310.8	290.7	296.3	385.4
Binat. non-wage services						34.0	20.7	31.4	56.0	141.7	221.2	225.5
Resource Balance	-248.6	-209.9	-138.2	74.7	-151.1	-216.1	-359.1	-423.6	-296.1	-161.2	241.0	65.5
Net Factor Services	94.4	136.7	113.5	55.4	30.7	-3.3	-93.1	-107.2	-67.3	-84.1	-28.8	-7.2
Interest Payments	77.4	78.7	79.9	64.1	78.3	91.6	167.3	117.4	138.0	138.0	112.5	101.1
Net binational wages	100.2	122.8	81.2	66.2	47.2	9.5	13.7	22.6	40.8	23.7	39.1	33.8
Other factor receipts (net)	71.6	92.6	112.2	53.3	61.8	78.8	60.5	-12.4	29.9	30.2	44.6	60.1
Transfers	4.5	5.7	5.0	6.2	9.3	7.5	11.1	27.0	35.2	35.2	23.9	32.9
Current Account	-149.7	-67.5	-19.7	136.3	-111.1	-211.9	-441.1	-503.8	-328.2	-210.1	236.1	91.2
Capital Account	275.5	220.9	177.9	106.3	109.3	96.1	214.7	84.5	-114.1	-190.0	-5.9	-42.2
Direct Investment	29.8	21.8	28.8	5.1	0.9	8.3	31.0	8.5	5.0	0.0	0.0	73.1
Net MLT Flows	148.1	133.9	225.7	281.3	209.5	83.3	128.2	-80.1	-77.3	-63.5	46.5	-424.3
Disbursements 3/	239.7	225.0	323.9	344.1	299.1	216.2	301.3	192.8	146.6	160.4	539.1	142.0
Amortizations 3/ 4/	91.6	91.1	98.2	62.8	89.6	132.9	173.1	272.9	223.9	223.9	492.6	566.3
Other Binational Flows 5/	28.6	25.4	21.6	5.7	9.6	69.9	45.8	70.9	10.0	-45.7	-42.0	2.1
Other Capital Flows 6/	69.0	39.8	-98.2	-185.8	-110.7	-65.4	9.7	85.2	-51.8	-80.8	-10.4	306.9
Others (E & O)	38.7	-108.2	-233.2	-282.1	-93.7	8.7	50.3	355.2	219.9	196.2	-92.7	38.7
Overall BOP Surplus	-164.5	-45.2	75.0	39.5	95.5	107.0	176.1	64.1	222.4	203.9	-137.5	-87.7
<> Intl. Reserves	-164.5	-45.2	75.0	39.5	95.5	91.1	92.6	-63.3	143.1	143.8	-136.7	-245.9
Arrears						15.9	83.5	127.4	79.3	60.1	-0.8	158.2
Amortizations						14.3	51.3	86.6	58.3	44.4	13.9	122.2
Interest						1.6	32.2	40.8	21.0	15.7	-14.7	36.0

Source: Central Bank of Paraguay and World Bank estimates.
1/ The first presentation of 1988 is with the old methodology of the Central Bank.
2/ Does not include binational non-factor services
3/ In 1989, this includes rescheduled Brazilian debt and principal and interest arrears.
4/ In 1990, the counterpart of the gift of the Brazilian debt reduction is included.
5/ Includes loans and arrears of Itaipu.
6/ In 1990, the gift of the Brazilian debt reduction is included; also includes short term flows.

27-Sep-91

Table 3.2: PARAGUAY - COMPOSITION OF REGISTERED EXPORTS (FOB), 1976-1989
(US$ Millions)

	1976	1977	1978	1979	1980	1981	1982	1983	1984	1985	1986	1987	1988	1989
In millions of current US$														
Merchandise Exports	177.7	274.6	261.2	299.5	308.9	295.3	327.6	263.9	329.9	302.3	198.4	332.1	486.5	911.9
Manufacturing	33.4	47.1	46.6	64.3	86.1	60.0	61.7	36.9	38.9	25.0	36.8	58.9	52.8	105.9
Primary Commodities	144.3	227.5	214.5	235.2	222.9	235.3	265.9	227.0	291.0	277.3	161.6	273.2	433.7	805.9
Non-Fuels	142.2	226.1	213.4	234.1	222.9	235.3	265.9	227.0	291.0	277.3	161.6	273.2	433.7	805.9
Food	96.4	137.0	109.0	131.8	107.3	98.8	139.7	136.9	149.2	131.1	109.5	183.0	240.5	593.3
Non-Food Agriculture	49.3	94.2	109.3	107.2	116.9	136.0	128.4	95.3	146.4	147.8	85.9	110.8	215.5	309.0
Metals and Minerals	0.0	0.0	0.0	0.0	0.0	0.7	0.0	0.1	0.0	0.0	0.0	0.7	0.7	0.2
Fuels	2.0	1.4	1.1	1.1	0.0	0.0	0.0	0.0	0.0	0.0	0.0	0.0	0.0	0.0
Memo items:														
Soybeans	34.1	58.8	41.6	81.3	45.3	52.5	91.0	88.5	101.6	106.3	45.8	125.0	157.8	390.2
Cotton	34.6	80.5	100.0	98.6	106.5	129.5	122.5	85.2	131.2	141.8	80.5	101.0	209.5	306.8
In millions of constant 1982 US$														
Merchandise Exports	280.5	327.2	311.9	571.0	437.3	311.2	327.6	280.9	297.0	295.7	212.0	403.2	371.3	819.8
Manufacturing	86.1	106.2	107.1	223.6	175.5	85.4	61.7	52.2	54.1	26.4	53.2	96.4	41.6	90.4
Primary Commodities	194.4	221.0	204.9	347.5	261.9	225.8	265.9	228.7	242.9	269.4	158.8	306.9	329.7	729.4
Non-Fuels	189.6	217.9	203.0	345.1	261.9	225.8	265.9	228.7	242.9	269.4	158.8	306.9	329.7	729.4
Food	128.9	132.3	104.7	193.1	138.2	101.8	139.7	149.2	146.5	139.4	127.0	236.8	167.4	539.4
Non-Food Agriculture	67.2	95.3	107.2	165.0	126.4	124.1	128.4	89.6	106.2	133.7	64.5	89.4	181.3	283.1
Metals and Minerals	0.0	0.0	0.0	0.0	0.0	0.2	0.0	0.1	0.0	0.0	0.0	0.8	4.1	4.3
Fuels	4.8	3.1	1.9	2.4	0.0	0.0	0.0	0.0	0.0	0.0	0.0	0.0	0.0	0.0
Memo items:														
Soybeans	51.5	57.9	41.9	118.3	64.8	53.4	91.0	97.8	93.9	104.9	64.1	179.4	108.5	380.8
Cotton	44.0	76.9	96.5	149.7	112.4	116.8	122.5	81.4	98.1	130.9	62.6	84.6	175.8	281.0
Deflators US$ (1982 = 1.00)														
Merchandise Exports	0.63	0.84	0.84	0.52	0.71	0.95	1.00	0.94	1.11	1.02	0.94	0.82	1.31	1.11
Manufacturing	0.39	0.44	0.44	0.29	0.49	0.70	1.00	0.71	0.72	0.95	0.69	0.61	1.27	1.17
Primary Commodities	0.74	1.03	1.05	0.68	0.85	1.04	1.00	0.99	1.20	1.03	1.02	0.89	1.32	1.10
Non-Fuels	0.75	1.04	1.05	0.68	0.85	1.04	1.00	0.99	1.20	1.03	1.02	0.89	1.32	1.10
Food	0.75	1.04	1.04	0.68	0.78	0.97	1.00	0.92	1.02	0.94	0.86	0.77	1.44	1.10
Non-Food Agriculture	0.73	0.99	1.02	0.65	0.92	1.10	1.00	1.06	1.38	1.11	1.33	1.24	1.19	1.09
Metals and Minerals	-	-	-	-	-	3.86	1.00	1.07	1.00	1.50	2.13	0.86	0.18	0.04
Fuels	0.42	0.47	0.60	0.45	-	0.80	-	-	-	-	-	2.67	-	-
Memo items:														
Soybeans	0.66	1.02	0.99	0.69	0.70	0.98	1.00	0.90	1.08	1.01	0.71	0.70	1.45	1.02
Cotton	0.79	1.05	1.04	0.66	0.95	1.11	1.00	1.05	1.34	1.08	1.29	1.19	1.19	1.09
Growth Rates (percent)														
Merchandise Exports	-	16.6	-4.7	83.1	-23.4	-28.8	5.3	-14.2	5.7	-0.4	-28.3	90.2	-7.9	120.8
Manufacturing	-	23.4	0.8	108.8	-21.5	-51.3	-27.7	-15.4	3.7	-51.3	101.8	81.1	-56.8	117.1
Primary Commodities	-	13.6	-7.3	69.6	-24.6	-13.8	17.7	-14.0	6.2	10.9	-41.0	93.3	7.4	121.2
Non-Fuels	-	14.9	-6.8	70.0	-24.1	-13.8	17.7	-14.0	6.2	10.9	-41.0	93.2	7.4	121.2
Food	-	0.1	-21.8	87.8	-24.7	-26.4	37.2	6.8	-1.8	-4.8	-8.9	86.4	-29.3	222.3
Non-Food Agriculture	-	50.5	12.4	54.0	-23.4	-1.8	3.5	-30.3	18.6	25.9	-51.7	38.6	102.7	56.2
Metals and Minerals	-	41.9	12.4	54.0	-23.4	-1.8	3.5	-30.3	18.6	25.9	-51.7	38.6	102.7	56.2
Fuels	-	-36.6	-39.3	30.9	-100.0	-	-100.0	-	-	-	-	-	-100.0	-
Memo items:														
Soybeans	-	12.4	-27.7	182.4	-45.3	-17.5	70.3	7.6	-4.1	11.8	-38.9	179.9	-39.5	250.8
Cotton	-	74.9	25.4	55.2	-24.9	3.9	4.8	-33.5	20.6	33.4	-52.2	35.2	107.8	59.8

Source: Central Bank of Paraguay and World Bank estimates.
16-Oct-91

Table 3.3: PARAGUAY - COMPOSITION OF REGISTERED IMPORTS (FOB), 1976-1989
(US$ Millions)

	1976	1977	1978	1979	1980	1981	1982	1983	1984	1985	1986	1987	1988	1989
In millions of current US$														
Merchandise Imports	194.1	300.0	307.4	420.7	455.1	461.2	509.3	411.2	441.6	358.4	379.9	517.5	366.5	714.4
Manufacturing	120.9	208.2	205.2	265.1	273.9	308.0	283.6	196.5	207.0	202.4	240.2	346.5	245.4	558.1
Capital Goods	69.9	129.6	118.4	144.5	148.2	178.4	150.3	110.8	115.9	95.9	132.6	184.0	130.3	273.9
Other manufactures	51.0	78.6	86.8	120.6	125.7	129.6	133.4	85.6	91.0	106.5	107.6	162.5	115.1	284.2
Primary Commodities	73.2	91.8	102.2	155.6	181.3	153.2	225.6	214.7	234.6	156.0	139.8	171.0	121.1	156.2
Non-Fuels	30.6	35.6	32.8	52.4	42.5	36.0	58.5	69.0	32.3	38.1	43.1	38.8	27.4	37.0
Food	12.2	11.2	10.4	14.2	17.5	7.7	13.8	24.0	7.1	16.8	10.5	5.7	4.0	7.9
Metals and Minerals	18.4	24.3	22.4	38.3	25.0	28.3	44.8	45.1	25.2	21.3	32.6	33.1	23.4	29.2
Fuels	42.6	56.2	69.4	103.2	138.8	117.2	167.1	145.6	202.3	117.9	96.7	132.2	93.6	119.2
In millions of constant US$														
Merchandise Imports	349.6	505.6	479.2	549.4	550.9	517.0	509.3	363.9	384.0	278.2	292.2	383.3	256.9	444.7
Manufacturing	200.6	331.8	269.8	336.2	365.5	341.0	283.6	183.9	180.9	177.9	188.2	247.9	166.1	330.9
Capital Goods	125.0	198.4	151.9	174.8	179.8	204.2	150.3	103.4	99.9	89.3	98.9	142.2	95.3	180.7
Other manufactures	75.6	133.4	117.9	161.4	185.8	136.7	133.4	80.5	81.0	88.5	89.3	105.6	70.8	150.1
Primary Commodities	149.0	173.7	209.4	213.1	185.3	176.1	225.6	180.0	203.1	100.3	104.1	135.5	90.8	113.8
Non-Fuels	36.2	50.2	44.5	66.8	50.3	50.6	58.5	61.6	26.4	26.5	27.3	24.3	16.3	21.1
Food	14.4	12.4	14.7	17.5	18.9	19.2	13.8	21.9	5.8	11.1	6.2	4.0	2.7	4.8
Metals and Minerals	21.8	37.8	29.8	49.3	31.4	31.4	44.8	39.7	20.7	15.4	21.1	20.3	13.6	16.3
Fuels	112.7	123.5	164.9	146.3	135.0	125.5	167.1	118.4	176.7	73.8	76.8	111.2	74.5	92.6
Deflators (1982=100)														
Merchandise Imports	0.56	0.59	0.64	0.77	0.83	0.89	1.00	1.13	1.15	1.29	1.30	1.35	1.43	1.61
Manufacturing	0.60	0.63	0.76	0.79	0.75	0.90	1.00	1.07	1.14	1.14	1.28	1.40	1.48	1.69
Capital Goods	0.56	0.65	0.78	0.83	0.82	0.87	1.00	1.07	1.16	1.07	1.34	1.29	1.37	1.52
Other manufactures	0.67	0.59	0.74	0.75	0.68	0.95	1.00	1.06	1.12	1.20	1.20	1.54	1.63	1.89
Primary Commodities	0.49	0.53	0.49	0.73	0.98	0.87	1.00	1.19	1.16	1.55	1.34	1.26	1.33	1.37
Non-Fuels	0.84	0.71	0.74	0.78	0.85	0.71	1.00	1.12	1.22	1.44	1.58	1.59	1.69	1.75
Food	0.84	0.90	0.71	0.81	0.93	0.40	1.00	1.10	1.23	1.51	1.67	1.41	1.49	1.63
Metals and Minerals	0.85	0.64	0.75	0.78	0.80	0.90	1.00	1.14	1.22	1.39	1.55	1.63	1.72	1.79
Fuels	0.38	0.46	0.42	0.71	1.03	0.93	1.00	1.23	1.15	1.60	1.26	1.19	1.26	1.29
Growth Rates														
Merchandise Imports	-	44.6	-5.2	14.7	0.3	-6.1	-1.5	-28.5	5.5	-27.5	5.1	31.2	-33.0	73.1
Manufacturing	-	65.4	-18.7	24.6	8.7	-6.7	-16.8	-35.2	-1.6	-1.7	5.8	31.7	-33.0	99.2
Capital Goods	-	58.7	-23.5	15.1	2.8	13.6	-26.4	-31.2	-3.4	-10.5	10.7	43.8	-33.0	89.6
Other manufactures	-	76.5	-11.6	36.9	15.1	-26.4	-2.5	-39.7	0.7	9.2	0.9	18.3	-33.0	112.0
Primary Commodities	-	16.6	20.5	1.8	-13.0	-5.0	28.1	-20.2	12.8	-50.6	3.7	30.2	-33.0	25.4
Non-Fuels	-	38.6	-11.4	50.2	-24.8	0.6	15.7	5.2	-57.1	0.3	3.2	-11.1	-33.0	29.9
Food	-	-14.1	18.3	19.6	7.6	1.5	-28.2	58.9	-73.7	93.3	-43.9	-35.5	-33.0	79.2
Metals and Minerals	-	73.6	-21.1	65.3	-36.3	0.0	42.5	-11.3	-48.0	-25.6	37.2	-3.8	-33.0	20.1
Fuels	-	9.6	33.5	-11.3	-7.7	-7.1	33.1	-29.1	49.1	-58.2	3.9	44.8	-33.0	24.4

Source: Central Bank of Paraguay and World Bank estimates.
16-Oct-91

Table 3.4: PARAGUAY - AVERAGE MARKET EXCHANGE RATE, 1980-1990
(Guaranies per US$)

	1980	1981	1982	1983	1984	1985	1986	1987	1988	1989	1990
January	138	136	161	255	339	402	773	716	892	1070	1290
February	138	138	155	238	335	408	831	710	893	1076	1255
March	138	137	156	236	366	435	790	719	885	1005	1217
April	137	136	158	242	368	532	771	751	888	1013	1237
May	136	138	165	270	365	551	732	796	889	1069	1255
June	136	140	182	311	404	609	706	801	892	1144	1212
July	135	144	192	333	405	663	671	801	911	1238	1201
August	135	148	227	423	404	848	651	806	924	1267	1198
September	134	158	257	422	425	786	623	848	954	1190	1196
October	135	165	282	390	403	723	598	885	970	1185	1201
November	135	170	257	343	399	655	612	890	1013	1204	1230
December	135	173	239	334	382	652	651	870	1026	1220	1270
Quarter I	138	137	157	243	347	415	798	715	890	1050	1254
Quarter II	136	138	168	274	379	564	736	783	890	1075	1235
Quarter III	135	150	225	393	411	766	648	818	930	1232	1198
Quarter IV	135	169	259	356	395	677	620	882	1003	1203	1234
Semester I	137	138	163	259	363	490	767	749	890	1063	1244
Semester II	135	160	242	374	403	721	634	850	966	1217	1216
Annual	136	149	203	316	383	605	701	799	928	1140	1230

Source: Foro de Economia, Centro Paraguayo de Estudios Sociologicos and Banco Central.
23-Sep-91

Table 4.1: PARAGUAY: EXTERNAL DEBT AND DEBT SERVICE, 1975 - 1990
(US$ Millions)

	1975	1980	1981	1982	1983	1984	1985	1986	1987	1988	1989	1990
Long Term Debt	228	784	842	1070	1273	1358	1638	1912	2253	2121	2124	1755
Official sources	139	405	455	557	706	789	1043	1258	1324	1256	1611	1205
o/w IBRD	8	68	99	147	168	179	248	315	372	315	282	279
Private sources	89	379	387	513	567	569	595	654	929	865	513	550
Short Term Debt	0	174	308	226	141	112	178	174	268	231	261	376
Interest Arrears					8	14	2	32	41	16	29	22
Total Debt	228	958	1150	1296	1414	1470	1816	2086	2521	2352	2385	2131
Long Term Debt Service Payments	30	125	131	106	101	137	158	222	225	296	142	196
Amortizations	22	80	95	61	53	76	78	132	131	183	77	121
Interest	9	45	36	45	48	61	80	90	94	113	65	75
Interest on Short Term Debt	0	22	40	36	1	1	0	1	15	20	10	13
Total Debt Service Payments	30	147	171	142	102	137	159	223	240	316	152	209
Debt Ratios												
Long Term Debt/GDP	15	18	15	20	23	31	52	54	60	54	52	33
Long Term Debt/Exports G&NFS	89	100	109	135	172	180	244	242	279	183	138	126
Debt Service Ratio a/	12	19	22	18	14	18	24	28	30	27	10	15
Total Debt/GDP	15	22	20	24	25	33	57	59	68	60	58	40
Public and Publicly Guaranteed LT Debt (as % of Total Long Term Debt)	83	81	84	88	90	92	94	96	99	99	99	99

a/ Total Debt Service Payments as a percent of exports of goods and services.
Source: World Bank - Debtor Reporting System.
23-Sep-91

Table 5.1: PARAGUAY - CONSOLIDATED PUBLIC SECTOR BUDGET, 1970-1990
(Current Guaranies, Millions)

	1970	1971	1972	1973	1974	1975	1976	1977	1978	1979
Current Revenues	12658	13099	14226	17173	23699	26921	30630	39791	51382	53422
Transfers	1	0	0	0	0	0	0	0	0	0
Central Government	1	0	0	0	0	0	0	0	0	0
Decentralized Agencies	0	0	0	0	0	0	0	0	0	0
Municipalities	0	0	0	0	0	0	0	0	0	0
Public Enterprises	0	0	0	0	0	0	0	0	0	0
Current Expenditures	10196	11174	12650	13013	16256	19338	22000	25774	29833	38247
Personnel Services	5699	6522	6744	6920	8114	9722	11584	13533	16094	19796
Other Services	2458	2608	2949	3443	4948	5574	6030	6678	7740	10115
Domestic Interest	87	88	154	126	226	177	227	302	249	489
External Interest	546	545	748	682	743	866	922	1274	1394	2110
Taxes	0	0	0	0	0	59	68	211	91	324
Transfers	1406	1411	2055	1842	2225	2940	3169	3774	4265	5412
Current Savings	2463	1925	1576	4160	7443	7583	8630	14018	21549	28978
Transfers and Loans	0	0	0	0	0	0	0	0	0	0
Total Capital Expenditures	3317	3741	4410	4912	7145	17117	15604	16072	21534	23703
Real Investment	2425	3075	4026	4989	5910	9046	12923	13822	16549	21188
Change in Inventories	361	235	129	-226	653	1073	789	907	2248	510
Financial Investment	531	431	255	149	582	6998	1892	1342	2737	2006
Financing Needs	855	1816	2834	752	-298	9534	6974	2054	-15	-5276
External Financing	751	798	1361	1963	2873	9790	6123	6481	7874	2523
Internal Financing	104	1018	1473	-1211	-3171	-256	851	-4427	-7890	-7798

Source: Ministry of Planning and Bank estimates
18-Sep-91

Table 5.1: PARAGUAY - CONSOLIDATED PUBLIC SECTOR BUDGET, 1970-1990
(Current Guaranies, Millions)

	1980	1981	1982	1983	1984	1985	1986	1987	1988	1989	1990
Current Revenues	82091	95111	111332	116642	140298	195957	250528	364029	473609	795775	1224473
Transfers											
Central Government	0	0	0	109	0	0	0	0	0	0	0
Decentralized Agencies	0	0	0	109	0	0	0	0	0	0	0
Municipalities	0	0	0	0	0	0	0	0	0	0	0
Public Enterprises	0	0	0	0	0	0	0	0	0	0	0
Current Expenditures	52353	72539	90639	103862	119218	142149	189787	269468	330924	635747	859680
Personnel Services	26762	37447	45990	53696	57479	70770	82007	112528	146031	244664	385447
Other Services	14656	20843	17231	19140	16134	26072	38780	49918	57389	89246	124624
Domestic Interest	415	451	915	1511	2713	2415	2292	4862	6150	6637	9725
External Interest	2587	3281	6301	5885	8001	9875	15122	32089	39770	142924	133476
Taxes	346	365	284	111	1321	29	13906	20743	19128	43263	57076
Transfers	7586	10153	19919	23518	33570	32988	37679	49329	62456	109014	149331
Current Savings	29738	22573	20693	12889	21080	53808	60741	94561	142685	160028	364793
Transfers and Loans	0	0	0	-207	-307	-447	-1779	-342	1232	-188	-4406
Total Capital Expenditures	29428	43265	40231	56112	89652	90237	92673	167754	246590	315118	322367
Real Investment	24958	29751	33059	54239	71650	71086	76520	98349	197070	231210	287697
Change in Inventories	1007	1067	3651	-1355	7210	11645	5531	21540	4877	55856	810
Financial Investment	3463	12447	3521	3227	10792	7506	10622	47865	44643	28052	33860
Financing Needs	-310	20692	19538	43430	68879	36876	33711	73536	102673	155277	-38021
External Financing	3071	6286	7721	29000	45110	13221	49203	12577	76685	66859	96033
Internal Financing	-3381	14406	11817	14430	23769	23656	-15493	60959	25988	88418	-134054

Source: Ministry of Planning and Bank estimates
18-Sep-91

Table 5.2: PARAGUAY - CENTRAL ADMINISTRATION BUDGET, 1970-1990
(Current Guaranies, Millions)

	1970	1971	1972	1973	1974	1975	1976	1977	1978	1979
Current Revenues	8784	8909	9478	11590	16262	17894	19244	26379	34333	43629
Transfers	-518	-563	-518	-534	-814	-1118	-1140	-1235	-1491	-2151
Central Government	-518	-563	-518	-534	-814	-1118	-1140	-1235	-1491	-2151
Decentralized Agencies	0	0	0	0	0	0	0	0	0	0
Municipalities	0	0	0	0	0	0	0	0	0	0
Public Enterprises	0	0	0	0	0	0	0	0	0	0
Current Expenditures	6851	7120	8118	8974	11240	13294	14913	17318	20125	25042
Personnel Services	3700	3875	4344	4490	5096	6126	7326	8529	10125	12205
Other Services	1955	2021	2320	2816	4226	4853	4964	5448	6181	7900
Domestic Interest	47	45	45	56	51	55	60	53	57	51
External Interest	201	187	322	247	272	308	323	638	795	1255
Taxes	0	0	0	0	0	0	0	0	0	0
Transfers	948	992	1087	1365	1595	1953	2239	2650	2967	3632
Current Savings	1415	1226	842	2082	4208	3482	3191	7826	12717	16436
Transfers and Loans	-378	-292	-290	-366	-734	-497	-562	-769	-966	-1067
Total Capital Expenditures	1119	1359	1520	1580	2143	3714	6926	6478	8194	12345
Real Investment	907	1090	1297	1388	1964	3471	5874	6051	7768	11379
Change in Inventories	0	0	0	0	0	0	0	0	119	0
Financial Investment	212	269	223	192	179	244	1051	427	307	966
Financing Needs	82	425	968	-136	-1331	730	4296	-579	-3557	-3025
External Financing	284	261	289	302	313	1363	4421	2526	1968	1956
Internal Financing	-202	164	679	-438	-1644	-633	-125	-3105	-5525	-4981

Source: Ministry of Planning and Bank estimates
18-Sep-91

Table 5.2: PARAGUAY - CENTRAL ADMINISTRATION BUDGET, 1970-1990
(Current Guaranies, Millions)

	1980	1981	1982	1983	1984	1985	1986	1987	1988	1989	1990
Current Revenues	51592	59107	68430	65459	85248	110249	143065	202376	263546	523391	816138
		0									
Transfers	-2915	-3941	-4278	-4732	-4561	-4949	-7634	-8173	-11186	-29252	-30066
Central Government	-2915	-3941	-4278	-4732	-4561	-4949	-7634	-8173	-11186	-29252	-30066
Decentralized Agencies	0	0	0	0	0	0	0	0	0	0	0
Municipalities	0	0	0	0	0	0	0	0	0	0	0
Public Enterprises	0	0	0	0	0	0	0	0	0	0	0
Current Expenditures	35149	49050	59817	68476	76459	92422	110868	152385	184056	353978	521287
Personnel Services	15938	22474	28288	32439	34319	40780	47108	63817	78994	141040	245000
Other Services	12340	17717	13434	15034	13157	19488	29294	37257	42285	70001	105899
Domestic Interest	59	107	86	119	989	908	784	682	580	478	717
External Interest	1707	2040	2458	2587	3899	5839	8895	19192	23140	67509	63561
Taxes	0	0	0	0	0	0	0	0	0	0	0
Transfers	5106	6712	15552	18297	24095	25408	24787	31437	39056	74950	106109
Current Savings	13528	6117	4335	-7749	4228	12878	24563	41818	68304	140161	264785
Transfers and Loans	-2761	-1528	-921	-834	-1325	-1277	-3037	-3328	-1361	-5300	-15000
Total Capital Expenditures	12151	24448	14486	12797	34353	33400	10733	40367	63692	47412	73071
Real Investment	10694	14743	14034	12677	29273	29716	10283	27742	61935	43864	43308
Change in Inventories	0	0	0	0	0	0	383	0	0	0	763
Financial Investment	1457	9705	452	120	5080	3684	67	12626	1756	3548	29000
Financing Needs	1384	19858	11072	21380	31450	21798	-10793	1877	-3251	-87450	-176714
External Financing	2538	2211	278	6107	14727	10719	4169	2953	9791	25957	8334
Internal Financing	-1154	17647	10794	15273	16723	11080	-14962	-1076	-13042	-113407	-185049

Source: Ministry of Planning and Bank estimates
18-Sep-91

Table 5.3: PARAGUAY - DECENTRALIZED AGENCIES BUDGET, 1970-1990
(Current Guaranies, Millions)

	1970	1971	1972	1973	1974	1975	1976	1977	1978	1979
Current Revenues	1729	1514	1748	2186	2538	3612	4106	5032	6295	7860
Transfers	356	409	451	470	579	742	816	981	1338	1740
Central Government	356	409	451	470	579	726	797	956	1284	1728
Decentralized Agencies	0	0	0	0	0	0	6	17	0	0
Municipalities	0	0	0	0	0	0	13	8	10	12
Public Enterprises	0	0	0	0	0	16	0	0	44	0
Current Expenditures	1490	1668	1927	1971	2368	2911	3358	3883	4772	6597
Personnel Services	758	836	908	987	1181	1295	1593	1835	2221	2959
Other Services	368	458	526	531	596	560	733	849	1204	1853
Domestic Interest	10	14	25	24	22	27	24	25	28	41
External Interest	11	13	21	18	19	42	75	107	85	85
Taxes	0	0	0	0	0	59	68	33	13	14
Transfers	343	347	447	411	550	928	866	1034	1222	1645
Current Savings	595	255	272	685	749	1443	1565	2131	2860	3002
Transfers and Loans	99	62	71	60	12	17	46	89	49	2
Total Capital Expenditures	576	326	295	259	822	1841	1828	1594	1301	2505
Real Investment	329	200	214	198	330	313	548	650	240	1451
Change in Inventories	-19	27	25	71	108	121	319	84	387	85
Financial Investment	266	99	56	-10	384	1407	961	860	674	969
Financing Needs	-118	9	-48	-486	61	381	217	-625	-1609	-499
External Financing	206	39	44	47	296	382	314	-62	-111	-122
Internal Financing	-324	-30	-92	-533	-235	-1	-97	-563	-1498	-377

Source: Ministry of Planning and Bank estimates
18-Sep-91

Table 5.3: PARAGUAY - DECENTRALIZED AGENCIES BUDGET, 1970-1990
(Current Guaranies, Millions)

	1980	1981	1982	1983	1984	1985	1986	1987	1988	1989	1990
Current Revenues	10396	12845	14061	17275	16922	20102	29839	44129	59852	85955	97560
Transfers	2114	2923	3268	3622	3472	3806	5519	6748	8689	15355	15403
Central Government	2072	2880	3232	3622	3463	3749	5459	6674	8600	15149	15403
Decentralized Agencies	25	24	0	0	9	9	12	0	0	0	0
Municipalities	17	19	27	0	0	48	48	61	75	137	0
Public Enterprises	0	0	9	0	0	0	0	13	15	69	0
Current Expenditures	7911	10377	12844	14722	18057	20932	29788	38857	48977	84515	84058
Personnel Services	3590	4661	5712	6489	6761	8546	9573	11165	13978	25065	32299
Other Services	1810	2431	2859	3158	1894	5189	7910	9728	11724	14948	13171
Domestic Interest	44	59	58	65	44	38	73	50	162	230	91
External Interest	82	77	70	91	68	163	197	296	213	11767	117
Taxes	21	40	48	50	314	29	90	717	872	217	0
Transfers	2364	3110	4097	4869	8977	6966	11946	16900	22028	32290	38380
Current Savings	4599	5391	4486	6175	2337	2976	5570	12020	19564	16795	28905
Transfers and Loans	1706	140	97	82	276	9	235	-734	595	3312	6000
Total Capital Expenditures	3370	4304	4804	5359	3434	4638	7229	9127	11690	21364	24497
Real Investment	1480	1852	2378	2914	1870	1967	1874	4419	3937	7953	19261
Change in Inventories	220	1	7	0	-246	100	786	281	431	3554	0
Financial Investment	1670	2451	2419	2445	1811	2571	4569	4428	7322	9857	5236
Financing Needs	-2936	-1227	222	-897	821	1653	1424	-2159	-8469	1257	-10408
External Financing	-60	-149	-61	574	213	144	1111	1361	311	-1568	8856
Internal Financing	-2876	-1078	283	-1471	608	1509	313	-3520	-8780	2825	-19264

Source: Ministry of Planning and Bank estimates
18-Sep-91

Table 5.4: PARAGUAY - MUNICIPALITIES BUDGET, 1970-1990
(Current Guaranies, Millions)

	1970	1971	1972	1973	1974	1975	1976	1977	1978	1979
Current Revenues	449	479	501	534	598	664	964	1282	1541	1933
Transfers	6	13	0	0	3	182	73	57	-10	48
Central Government	6	13	0	0	3	182	86	66	0	60
Decentralized Agencies	0	0	0	0	0	0	0	0	0	0
Municipalities	0	0	0	0	0	0	-13	-8	-10	-12
Public Enterprises	0	0	0	0	0	0	0	0	0	0
Current Expenditures	378	393	377	399	453	544	828	1077	1141	1277
Personnel Services	240	251	265	287	306	359	469	632	746	874
Other Services	135	129	103	96	126	162	333	381	356	362
Domestic Interest	0	0	0	4	6	6	18	37	26	28
External Interest	0	1	0	0	1	1	1	2	0	0
Taxes	0	0	0	0	0	0	0	0	0	0
Transfers	3	12	9	12	14	17	6	25	14	13
Current Savings	77	99	124	135	148	302	209	262	389	704
Transfers and Loans	-1	0	0	-5	0	-4	0	0	57	0
Total Capital Expenditures	68	74	70	188	222	353	415	511	713	835
Real Investment	91	88	112	231	274	368	572	474	817	907
Change in Inventories	0	29	0	0	0	0	0	0	0	0
Financial Investment	-23	-43	-42	-43	-52	-15	-157	37	-105	-72
Financing Needs	-8	-25	-54	58	74	56	206	249	267	131
External Financing	-6	-9	-5	-4	28	-10	-3	-10	-13	-20
Internal Financing	-2	-16	-49	62	46	66	208	258	280	151

Source: Ministry of Planning and Bank estimates
18-Sep-91

- 172 -

Table 5.4: PARAGUAY - MUNICIPALITIES BUDGET, 1970-1990
(Current Guaranies, Millions)

	1980	1981	1982	1983	1984	1985	1986	1987	1988	1989	1990
Current Revenues	2633	3244	3832	4476	4901	5898	6986	9220	11334	13542	17500
Transfers	83	116	172	252	302	258	441	349	511	520	629
Central Government	100	135	200	252	302	306	489	410	586	656	629
Decentralized Agencies	0	0	0	0	0	0	0	0	0	0	0
Municipalities	-17	-19	-27	0	0	-48	-48	-61	-75	-137	0
Public Enterprises	0	0	0	0	0	0	0	0	0	0	0
Current Expenditures	1746	2344	2714	3173	3255	4180	4893	6109	7491	10456	12630
Personnel Services	1129	1548	1834	2070	1992	2553	2879	3695	4755	7399	8020
Other Services	501	605	707	817	940	1161	1425	1831	1981	2157	3340
Domestic Interest	49	43	25	126	59	79	118	150	152	150	250
External Interest	0	0	0	0	0	0	9	4	12	8	20
Taxes	0	0	0	0	0	0	0	0	0	0	0
Transfers	67	148	148	160	263	386	463	430	591	743	1000
Current Savings	970	1016	1290	1556	1948	1977	2534	3460	4354	3605	5499
Transfers and Loans	199	0	0	0	0	0	0	0	0	0	0
Total Capital Expenditures	1357	1516	1718	1826	2431	3411	3447	5175	6647	7490	11600
Real Investment	1542	1618	1914	2044	2628	3697	3667	5547	7157	7462	12100
Change in Inventories	0	0	0	0	0	0	33	0	0	0	0
Financial Investment	-185	-102	-197	-218	-197	-286	-253	-373	-510	28	-500
Financing Needs	189	501	428	271	483	1435	913	1715	2294	3885	6101
External Financing	0	105	0	0	69	0	-12	-36	-67	-90	0
Internal Financing	189	396	428	271	414	1435	925	1751	2361	3975	6101

Source: Ministry of Planning and Bank estimates
18-Sep-91

Table 5.5: PARAGUAY - PUBLIC ENTERPRISES BUDGET, 1970-1990
(Current Guaranies, Millions)

	1970	1971	1972	1973	1974	1975	1976	1977	1978	1979
Current Revenues	1696	2197	2499	2863	4301	4751	6316	7098	9214	13803
Transfers	157	141	67	64	232	194	251	196	164	363
Central Government	157	141	67	64	232	210	257	213	208	363
Decentralized Agencies	0	0	0	0	0	0	-6	-17	0	0
Municipalities	0	0	0	0	0	0	0	0	0	0
Public Enterprises	0	0	0	0	0	-16	0	0	-44	0
Current Expenditures	1477	1993	2228	1669	2195	2589	2902	3495	3795	5331
Personnel Services	1001	1560	1227	1156	1531	1943	2196	2538	3001	3759
Other Services	0	0	0	0	0	0	0	0	0	0
Domestic Interest	30	29	84	42	147	89	125	186	139	369
External Interest	334	344	405	417	451	515	523	528	514	770
Taxes	0	0	0	0	0	0	0	178	79	310
Transfers	112	60	512	54	66	42	58	65	62	122
Current Savings	376	345	338	1258	2338	2357	3666	3799	5583	8836
Transfers and Loans	280	230	219	311	722	484	516	680	859	1065
Total Capital Expenditures	1554	1982	2525	2885	3958	11208	6436	7489	11326	8018
Real Investment	1098	1697	2403	3172	3342	4894	5929	6647	7723	7451
Change in Inventories	380	179	104	-297	545	952	470	823	1743	425
Financial Investment	76	106	18	10	71	5363	37	19	1860	142
Financing Needs	899	1407	1968	1316	898	8368	2255	3010	4884	-1882
External Financing	267	507	1033	1618	2236	8056	1390	4027	6030	708
Internal Financing	632	900	935	-302	-1338	312	865	-1017	-1147	-2591
Memorandum:										
Sales of Goods & Services	3655	4129	4479	5060	6999	8471	9573	13447	16667	23988
Value of Inputs	1959	1932	1980	2197	2698	3719	3257	6348	7453	10185
Value Added	1696	2197	2499	2863	4301	4751	6316	7098	9214	13803

Source: Ministry of Planning and Bank estimates
18-Sep-91

- 174 -

Table 5.5: PARAGUAY - PUBLIC ENTERPRISES BUDGET, 1970-1990
(Current Guaranies, Millions)

	1980	1981	1982	1983	1984	1985	1986	1987	1988	1989	1990
Current Revenues	17471	19916	25009	29431	33228	59708	70637	108305	138878	172888	293275
Transfers		0									
Central Government	718	902	838	967	787	884	1674	1077	1986	13377	14034
Decentralized Agencies	744	926	846	967	796	894	1686	1090	2001	13446	14034
Municipalities	-25	-24	0	0	-9	-9	-12	0	0	0	0
Public Enterprises	0	0	0	0	0	0	0	0	0	0	0
	0	0	-9	0	0	0	0	-13	-15	-69	0
Current Expenditures	7547	10768	15264	17491	21448	24615	44237	72118	90400	186798	241705
Personnel Services	6107	8764	10157	12699	14407	18891	22447	33851	48304	71160	100128
Other Services	6	91	230	132	143	234	151	1102	1399	2142	2214
Domestic Interest	263	243	746	1201	1621	1389	1318	3979	5255	5778	8667
External Interest	799	1164	3772	3207	4035	3873	6021	12598	16405	63640	69778
Taxes	325	325	237	61	1008	0	13816	20026	18256	43046	57076
Transfers	49	182	122	192	235	229	483	562	782	1032	3842
Current Savings	10641	10050	10582	12907	12567	35977	28074	37264	50464	-533	65604
Transfers and Loans	856	1388	824	545	742	820	1023	3719	1998	1800	4594
Total Capital Expenditures	12550	12997	19223	36130	49434	48788	71264	113085	164561	238852	213199
Real Investment	11242	11538	14732	36604	37880	35707	60697	60642	124040	171931	213028
Change in Inventories	786	1067	3644	-1355	7456	11545	4329	21259	4446	52302	47
Financial Investment	522	393	848	881	4099	1536	6238	31184	36075	14619	124
Financing Needs	1052	1560	7816	22677	36125	11990	42167	72103	112099	237585	143001
External Financing	593	4119	7503	22319	30101	2358	43936	8299	66650	42560	78843
Internal Financing	460	-2559	313	358	6024	9632	-1769	63804	45449	195025	64158
Memorandum:											
Sales of Goods & Services	29614	33715	72617	76932	97669	157564	168935	248001	329102	416541	609088
Value of Inputs	13209	15355	49409	48847	70701	109379	111701	152541	214670	273782	332825
Value Added	16406	18360	23208	28084	26968	48185	57234	95460	114432	142759	276263

Source: Ministry of Planning and Bank estimates
18-Sep-91

Table 5.6: PARAGUAY - EXECUTED BUDGET OF PUBLIC ENTERPRISES, 1980-1990
(Current Guaranies, Millions)

	1980	1981	1982	1983	1984	1985	1986	1987	1988	1989	1990
CURRENT ACCOUNT	11498	11437	11407	13453	13309	36797	29097	40982	52462	2407	70373
CURRENT REVENUES	19045	22206	26671	30944	34757	61413	73334	113100	142862	189196	312005
Value Added	16406	18360	23208	28084	26968	48185	57234	95460	114432	142759	276263
Sales of Goods & Services	29614	33715	72617	76932	97669	157564	168935	248001	329102	416541	609088
(Cost of Inputs)	13209	15355	49409	48847	70701	109379	111701	152541	214670	273782	332825
Other	1065	1556	1801	1347	6260	11524	13404	12845	24446	30129	17012
Current Transfers	718	902	838	967	787	884	1674	1077	1986	14508	14136
Capital Transfers & Loans	856	1388	824	545	742	820	1023	3719	1998	1800	4594
CURRENT EXPENDITURES	7547	10768	15264	17491	21448	24615	44237	72118	90400	186789	241632
Operating Costs	7173	10261	14905	17238	20205	24387	29937	51530	71363	142711	180715
Wages & Salaries	5787	8103	9330	11616	12976	17379	20789	30309	42980	63478	89792
Social Security Contrib.	320	661	828	1082	1431	1512	1658	3543	5323	7682	10336
Rent	6	91	230	132	143	234	151	1102	1399	2142	2214
Interest on Internal Debt	263	243	746	1201	1621	1389	1318	3979	5255	5778	8667
Interest on External Debt	799	1164	3772	3207	4035	3873	6021	12598	16405	63640	69778
Taxes & Transfers	374	507	359	253	1242	229	14300	20588	19038	44077	60918
Taxes	325	325	237	61	1008	0	13816	20026	18256	43046	57076
Transfers	49	182	122	192	235	229	483	562	782	1032	3842
CAPITAL ACCOUNT	12550	12997	19223	36130	49434	48788	71264	113085	164561	238852	213199
Real Investment	11242	11538	14732	36604	37880	35707	60697	60642	124040	171931	213028
Change in Inventories	786	1067	3644	-1355	7456	11545	4329	21259	4446	52302	47
Financial Investment	582	430	940	1208	4281	1669	6735	32405	36690	14685	424
Net Loans	0	0	-43	64	0	66	19	0	0	0	0
Capital Revenues	60	37	49	391	182	199	515	1221	615	67	300
BALANCE	-1052	-1560	-7816	-22677	-36125	-11990	-42167	-72103	-112099	-236445	-142826

Source: Luis A. Campos & Ricardo Canese, "El Sector Publico en el Paraguay" and Ministry of Finance.
18-Sep-91

Table 5.7: PARAGUAY - EXECUTED BUDGET OF ADMINISTRACION NACIONAL DE ELECTRICIDAD (ANDE), 1980-1990
(Current Guaranies, Millions)

	1980	1981	1982	1983	1984	1985	1986	1987	1988	1989	1990
CURRENT ACCOUNT	5124	5448	5599	5699	7708	10481	10855	15986	22833	32155	31589
CURRENT REVENUES	6812	7666	8143	8631	11174	14582	16454	24047	35607	55881	67857
Value Added	6158	6584	7164	7808	10689	12900	12708	20093	29832	38048	57957
Sales of Goods & Services	7496	8013	8577	9445	11865	16430	17578	30215	41561	75204	116134
(Cost of Inputs)	1338	1429	1413	1637	1176	3529	4869	10122	11729	37155	58177
Other	654	1081	979	823	485	1682	3745	3954	5775	17832	9900
Current Transfers	0	0	0	0	0	0	0	0	0	0	0
Capital Transfers & Loans	0	0	0	0	0	0	0	0	0	0	0
CURRENT EXPENDITURES	1688	2218	2544	2932	3466	4102	5599	8061	12774	23726	36269
Operating Costs	1641	2130	2448	2826	3386	4007	5467	7918	12584	23422	35868
Wages & Salaries	1323	1551	1799	1966	2219	2899	3723	5277	8823	13138	22825
Social Security Contrib.	82	176	200	226	306	414	525	794	1288	1852	3124
Rent	0	31	62	81	109	130	83	98	176	271	531
Interest on Internal Debt	1	1	0	1	1	1	0	0	0	1	0
Interest on External Debt	235	372	387	552	751	564	1136	1749	2296	8160	9387
Taxes & Transfers	47	88	97	107	80	95	131	143	190	304	401
Taxes	38	0	0	0	0	0	0	0	0	0	0
Transfers	9	88	97	107	80	95	131	143	190	304	401
CAPITAL ACCOUNT	3745	2431	2361	4405	15163	8251	20733	21910	38638	145284	195609
Real Investment	3853	2373	2361	4378	12432	7427	13904	15936	26559	133425	195609
Change in Inventories	-547	0	0	0	455	966	1836	-171	2037	11803	0
Financial Investment	463	102	64	69	2276	26	5461	6506	10042	56	0
Net Loans	0	0	-43	-21	0	0	0	0	0	0	0
Capital Revenues	24	44	22	22	0	168	468	360	0	0	0
BALANCE	1378	3017	3238	1294	-7455	2230	-9878	-5924	-15805	-113129	-164020

Source: Luis A. Campos & Ricardo Canese, "El Sector Publico en el Paraguay" and Ministry of Finance.
18-Sep-91

Table 5.8: PARAGUAY - EXECUTED BUDGET OF ADMINISTRACION NACIONAL DE TELECOMUNICACIONES (ANTELCO), 1980-90
(Current Guaranies, Millions)

	1980	1981	1982	1983	1984	1985	1986	1987	1988	1989	1990
CURRENT ACCOUNT	3522	2757	3255	3750	3655	4219	3904	5887	9048	-7421	16751
CURRENT REVENUES	5372	6296	7315	8660	8638	11999	11792	18688	24343	31819	66303
Value Added	5173	6114	7109	8545	8527	11736	11669	18456	23638	30077	65432
Sales of Goods & Services	7235	8786	10121	11619	12969	18076	22328	33374	45300	67046	81035
(Cost of Inputs)	2062	2673	3012	3074	4442	6340	10659	14919	21662	36969	15604
Other	199	182	207	115	111	263	124	232	705	1742	871
Current Transfers	0	0	0	0	0	0	0	0	0	0	0
Capital Transfers & Loans	0	0	0	0	0	0	0	0	0	0	0
CURRENT EXPENDITURES	1851	3539	4060	4909	4983	7780	7888	12801	15295	39240	49552
Operating Costs	1836	3476	4060	4843	4929	7780	7763	12620	15071	38783	46341
Wages & Salaries	1325	2767	2766	3398	3723	5406	4987	7652	9544	19037	24522
Social Security Contrib.	66	261	339	413	450	671	641	884	1327	2325	3623
Rent	0	4	11	5	1	8	0	0	53	70	84
Interest on Internal Debt	24	24	135	189	358	149	37	684	218	405	607
Interest on External Debt	420	420	810	838	398	1547	2099	3400	3930	16947	17505
Taxes & Transfers	15	63	0	67	54	0	125	181	224	457	3211
Taxes	0	0	0	0	0	0	0	0	0	0	0
Transfers	15	63	0	67	54	0	125	181	224	457	3211
CAPITAL ACCOUNT	3432	2851	5494	3653	4712	3624	4612	14296	41392	17823	23689
Real Investment	2931	2631	2822	2177	2443	5522	3497	10288	37894	15776	23476
Change in Inventories	501	23	2281	1476	704	-1899	1115	2575	2745	991	0
Financial Investment	0	197	391	0	1565	0	0	1432	753	1056	213
Net Loans	0	0	0	0	0	0	0	0	0	0	0
Capital Revenues	0	0	0	0	0	0	0	0	0	0	0
BALANCE	90	-94	-2239	98	-1057	595	-708	-8409	-32344	-25244	-6939

Source: Luis A. Campos & Ricardo Canese, "El Sector Publico en el Paraguay" and Ministry of Finance.
18-Sep-91

Table 5.9: PARAGUAY - EXECUTED BUDGET OF CORPORACION DE OBRAS SANITARIAS (CORPOSANA), 1980-1990
(Current Guaranies, Millions)

	1980	1981	1982	1983	1984	1985	1986	1987	1988	1989	1990
CURRENT ACCOUNT	1130	1322	1371	588	1028	1386	1593	2328	3501	3580	6602
CURRENT REVENUES	1840	2165	2116	1553	2414	3238	3745	5157	6554	9546	14053
Value Added	1181	1079	1406	1204	1660	2383	2433	4116	5213	7560	11003
Sales of Goods & Services	1764	1960	2081	2021	2882	4043	4181	6470	8213	11363	15967
(Cost of Inputs)	583	880	675	817	1222	1660	1749	2353	3001	3803	4964
Other	11	47	118	51	12	35	289	50	110	186	170
Current Transfers	0	0	0	0	0	0	0	0	0	0	0
Capital Transfers & Loans	648	1038	593	298	742	820	1023	991	1232	1800	2879
CURRENT EXPENDITURES	709	843	745	964	1386	1852	2152	2829	3054	5966	7451
Operating Costs	707	838	744	962	1383	1852	2149	2821	3048	5955	7433
Wages & Salaries	502	624	564	766	982	1033	1193	1249	1993	3226	4805
Social Security Contrib.	55	69	71	98	134	125	142	295	269	394	614
Rent	0	6	6	5	7	8	7	0	17	24	26
Interest on Internal Debt	25	80	21	20	24	33	25	79	143	111	167
Interest on External Debt	126	59	81	73	236	653	782	1198	626	2199	1820
Taxes & Transfers	2	5	2	2	3	0	3	8	6	12	18
Taxes	0	0	0	0	0	0	0	0	0	0	0
Transfers	2	5	2	2	3	0	3	8	6	12	18
CAPITAL ACCOUNT	1428	3157	3739	2206	3513	1545	7134	7053	7047	11373	27605
Real Investment	1371	2948	3694	2206	3314	1500	7276	7687	6966	11572	27605
Change in Inventories	84	201	63	0	177	47	-125	190	692	-281	0
Financial Investment	0	8	0	0	22	12	23	32	4	91	0
Net Loans	0	0	0	0	0	0	0	0	0	0	0
Capital Revenues	28	0	18	0	0	13	40	855	615	9	0
BALANCE	-298	-1836	-2368	-1618	-2485	-159	-5541	-4725	-3546	-7793	-21003

Source: Luis A. Campos & Ricardo Canese, "El Sector Publico en el Paraguay" and Ministry of Finance.
18-Sep-91

Table 5.10: PARAGUAY - EXECUTED BUDGET OF FLOTA MERCANTE DEL ESTADO (FLOMERES), 1980-1990
(Current Guaranies, Millions)

	1980	1981	1982	1983	1984	1985	1986	1987	1988	1989	1990
CURRENT ACCOUNT	98	42	144	72	491	1422	959	1562	1482	-4078	2677
CURRENT REVENUES	371	321	431	429	922	2534	2080	3119	3720	1848	9584
Value Added	229	172	187	292	922	2523	2069	3058	3720	1848	2848
Sales of Goods & Services	622	606	601	714	1734	4567	4552	6332	7317	9880	10656
(Cost of Inputs)	392	434	414	422	811	2044	2482	3274	3598	8031	7808
Other	8	11	24	2	0	11	11	62	0	0	245
Current Transfers	0	0	0	0	0	0	0	0	0	0	6491
Capital Transfers & Loans	134	137	219	134	0	0	0	0	0	0	0
CURRENT EXPENDITURES	274	279	287	357	432	1112	1122	1558	2238	5926	6907
Operating Costs	273	278	286	357	432	1083	1085	1552	2230	5918	6850
Wages & Salaries	225	228	248	328	400	420	501	604	972	1665	2103
Social Security Contrib.	23	25	27	29	30	38	45	56	96	144	193
Rent	0	0	0	0	0	0	0	0	0	0	0
Interest on Internal Debt	26	25	10	0	1	0	0	99	72	104	155
Interest on External Debt	0	0	0	0	0	626	540	794	1091	4005	4398
Taxes & Transfers	1	1	1	0	0	29	37	6	8	8	57
Taxes	0	0	0	0	0	0	0	0	0	0	0
Transfers	1	1	1	0	0	29	37	6	8	8	57
CAPITAL ACCOUNT	121	151	1273	4025	4228	1543	1315	917	1981	319	409
Real Investment	111	151	1311	4098	4245	1519	1293	888	1941	47	409
Change in Inventories	10	0	-38	-42	-17	24	22	29	40	276	0
Financial Investment	0	0	0	0	0	0	0	0	0	0	0
Net Loans	0	0	0	0	0	0	0	0	0	0	0
Capital Revenues	0	0	0	31	0	0	0	0	0	4	0
BALANCE	-24	-109	-1129	-3954	-3738	-121	-356	644	-499	-4396	2268

Source: Luis A. Campos & Ricardo Canese, "El Sector Publico en el Paraguay" and Ministry of Finance.
18-Sep-91

Table 5.11: PARAGUAY - EXECUTED BUDGET OF INDUSTRIA NACIONAL DEL CEMENTO (INC), 1980-1990
(Current Guaranies, Millions)

	1980	1981	1982	1983	1984	1985	1986	1987	1988	1989	1990
CURRENT ACCOUNT	831	997	-1040	545	250	209	-1079	-1396	-4148	-17661	-17323
CURRENT REVENUES	1153	1383	369	1305	3556	1254	1878	7069	8857	14434	20762
Value Added	1147	1376	341	1305	3425	1254	1515	6437	7205	13877	20762
Sales of Goods & Services	2696	3235	3633	3477	4949	6113	9818	13195	18335	22784	33723
(Cost of Inputs)	1549	1859	3292	2172	1524	4859	8303	6758	11130	8907	12961
Other	6	7	28	0	131	0	363	632	1652	558	0
Current Transfers	0	0	0	0	0	0	0	0	0	0	0
Capital Transfers & Loans	0	0	0	0	0	0	0	0	0	0	0
CURRENT EXPENDITURES	322	386	1409	760	3306	1045	2956	8465	13004	32096	38085
Operating Costs	315	379	1398	754	3306	1040	2942	8448	12984	32078	38085
Wages & Salaries	288	346	518	469	586	783	1151	1814	2358	4150	5132
Social Security Contrib.	18	22	52	75	90	76	147	215	280	347	409
Rent	4	5	26	15	0	12	20	29	40	18	0
Interest on Internal Debt	5	7	137	195	234	64	186	2389	3725	4765	7147
Interest on External Debt	0	0	665	0	2396	106	1438	4001	6582	22799	25397
Taxes & Transfers	7	8	11	6	0	5	14	17	20	18	0
Taxes	0	0	0	0	0	0	0	0	0	0	0
Transfers	7	8	11	6	0	5	14	17	20	18	0
CAPITAL ACCOUNT	824	989	430	20116	12966	16438	28732	12212	34578	8709	889
Real Investment	693	832	343	20032	12854	14095	28885	12104	32354	2240	829
Change in Inventories	131	157	0	0	191	2204	-153	115	2224	6441	0
Financial Investment	0	0	87	0	103	139	0	0	0	28	60
Net Loans	0	0	0	84	0	0	0	0	0	0	0
Capital Revenues	0	0	0	0	182	0	0	7	0	0	0
BALANCE	7	8	-1470	-19571	-12716	-16229	-29810	-13608	-38725	-26370	-18212

Source: Luis A. Campos & Ricardo Canese, "El Sector Publico en el Paraguay" and Ministry of Finance.
18-Sep-91

Table 5.12: PARAGUAY - EXECUTED BUDGET OF LINEAS AEREAS PARAGUAYAS (LAP), 1980-1990
(Current Guaranies, Millions)

	1980	1981	1982	1983	1984	1985	1986	1987	1988	1989	1990
CURRENT ACCOUNT	19	388	180	803	-324	947	-1675	1075	-2210	2250	-12345
CURRENT REVENUES	902	1694	2306	4018	2293	5081	3972	11590	11577	18910	10732
Value Added	490	1042	1886	3497	-1789	-2622	-939	5589	-4304	2547	1159
Sales of Goods & Services	5566	6401	16456	17281	21874	41660	29667	45563	50082	71011	88743
(Cost of Inputs)	5077	5359	14571	13784	23664	44282	30606	39973	54386	68464	87584
Other	0	0	0	0	3641	7295	4332	3273	14327	3863	1861
Current Transfers	339	440	408	408	441	408	580	0	894	12500	6512
Capital Transfers & Loans	74	212	12	113	0	0	0	2728	660	0	1200
CURRENT EXPENDITURES	883	1305	2126	3215	2617	4134	5647	10515	13787	16660	23077
Operating Costs	883	1305	2126	3215	2617	4134	5647	10515	13787	16660	23077
Wages & Salaries	804	964	1254	2349	2000	3342	4879	7565	9658	11514	16490
Social Security Contrib.	0	0	0	0	0	0	0	915	1107	1550	2023
Rent	0	39	99	0	0	0	0	922	1027	1573	1390
Interest on Internal Debt	79	13	38	290	617	793	768	135	246	371	557
Interest on External Debt	0	289	735	576	0	0	0	979	1749	1652	2617
Taxes & Transfers	0	0	0	0	0	0	0	0	0	0	0
Taxes	0	0	0	0	0	0	0	0	0	0	0
Transfers	0	0	0	0	0	0	0	0	0	0	0
CAPITAL ACCOUNT	724	427	270	2385	1688	3721	1606	11618	8944	14173	5300
Real Investment	723	427	243	2189	1392	3511	1549	11222	8481	7681	5300
Change in Inventories	1	0	37	197	296	210	57	396	463	6492	0
Financial Investment	0	0	0	0	0	0	0	0	0	0	0
Net Loans	0	0	0	0	0	0	0	0	0	0	0
Capital Revenues	0	0	10	0	0	0	0	0	0	0	0
BALANCE	-704	-38	-90	-1582	-2012	-2774	-3280	-10544	-11154	-11923	-17645

Source: Luis A. Campos & Ricardo Canese, "El Sector Publico en el Paraguay" and Ministry of Finance.
18-Sep-91

Table 5.13: PARAGUAY - EXECUTED BUDGET OF SIDEPAR AND ACEPAR, 1980-1990 a/
(Current Guaranies, Millions)

	1980	1981	1982	1983	1984	1985	1986	1987	1988	1989	1990
CURRENT ACCOUNT	123	117	104	203	0	0	494	301	319	-3549	-593
CURRENT REVENUES	136	128	138	254	53	61	560	1117	2550	7697	11034
Value Added	-1	-2	-6	-5	-5	-5	-7	703	2105	5997	8620
Sales of Goods & Services	0	0	0	0	0	0	0	3765	11474	15389	22001
(Cost of Inputs)	1	2	6	5	5	5	7	3062	9369	9392	13381
Other	0	0	0	0	0	0	0	0	0	823	766
Current Transfers	137	131	143	259	58	66	567	414	445	877	1133
Capital Transfers & Loans	0	0	0	0	0	0	0	0	0	0	515
CURRENT EXPENDITURES	12	11	34	51	53	61	66	816	2231	11247	11627
Operating Costs	12	11	34	51	53	61	66	816	2231	11202	11557
Wages & Salaries	12	11	33	51	51	58	66	598	1629	2576	2948
Social Security Contrib.	0	0	0	0	0	0	0	210	572	786	0
Rent	0	1	1	1	2	2	0	8	17	6	6
Interest on Internal Debt	0	0	0	0	0	0	0	0	0	0	0
Interest on External Debt	0	0	0	0	0	0	0	0	13	7834	8603
Taxes & Transfers	0	0	0	0	0	0	0	0	0	45	70
Taxes	0	0	0	0	0	0	0	0	0	45	70
Transfers	0	0	0	0	0	0	0	0	0	0	0
CAPITAL ACCOUNT	123	117	373	932	307	401	506	24303	24906	6012	1649
Real Investment	5	4	7	10	0	25	11	5	34	40	1611
Change in Inventories	0	0	0	0	0	0	0	0	0	0	0
Financial Investment	119	113	367	923	307	394	495	24298	24872	5987	38
Net Loans	0	0	0	0	0	0	0	0	0	0	0
Capital Revenues	0	0	0	0	0	17	0	0	0	15	0
BALANCE	0	0	-269	-729	-307	-401	-12	-24002	-24587	-9562	-2241

a/ Siderurgia Paraguaya and Aceros de Paraguay.
Source: Luis A. Campos & Ricardo Canese, "El Sector Publico en el Paraguay" and Ministry of Finance.
18-Sep-91

Table 5.14: PARAGUAY - EXECUTED BUDGET OF PETROPAR AND APAL, 1980-1990 a/
(Current Guaranies, Millions)

	1980 b/	1981	1982	1983	1984	1985	1986	1987	1988	1989	1990
CURRENT ACCOUNT	226	112	1604	1349	48	17347	13115	13983	20100	-3834	34167
CURRENT REVENUES	914	923	4079	3979	3411	19852	29378	37934	43469	45404	97074
Value Added	886	919	4026	3933	2109	17816	25008	33758	42348	39780	94429
Sales of Goods & Services	2537	2932	29294	30199	38985	63171	76604	103474	139495	143728	233960
(Cost of Inputs)	1651	2013	25268	26265	36876	45355	51597	69716	97147	103948	139531
Other	29	4	53	45	1302	2036	4370	4176	1121	5624	2645
Current Transfers	0	0	0	0	0	0	0	0	0	0	0
Capital Transfers & Loans	0	0	0	0	0	0	0	0	0	0	0
CURRENT EXPENDITURES	688	811	2474	2630	3363	2505	16263	23951	23370	49238	62908
Operating Costs	387	472	2226	2559	2258	2408	2278	3730	4792	4573	5043
Wages & Salaries	254	323	663	739	1279	1534	1867	2717	3604	3833	3979
Social Security Contrib.	47	63	89	185	370	156	97	130	311	367	395
Rent	0	0	6	10	7	17	17	16	37	28	39
Interest on Internal Debt	87	86	394	495	363	341	292	472	841	243	259
Interest on External Debt	0	0	1074	1130	240	360	6	395		102	371
Taxes & Transfers	301	339	249	71	1105	97	13985	20221	18578	44665	57865
Taxes	287	325	237	61	1008	0	13816	20022	18254	44373	57506
Transfers	14	14	12	10	97	97	169	198	324	292	359
CAPITAL ACCOUNT	1448	2088	4908	-1956	6453	12716	5737	19418	828	33407	479
Real Investment	888	1446	3569	944	824	1568	3456	1272	3634	701	479
Change in Inventories	560	642	1308	-2935	5629	10004	1530	18142	-3821	25417	0
Financial Investment	0	2	31	36	0	1078	731	4	1015	7289	0
Net Loans	0	0	0	0	0	66	19	0	0	0	0
Capital Revenues	0	1	0	0	0	0	0	0	0	0	0
BALANCE	-1222	-1976	-3304	3305	-6405	4631	7378	-5436	19271	-37241	33688

a/ Petroleos Paraguaya and Administracion Paraguaya de Alcoholes.
b/ 1980-1981 budgets exclude PETROPAR.
Source: Luis A. Campos & Ricardo Canese, "El Sector Publico en el Paraguay" and Ministry of Finance.
16-Oct-91

Table 5.15: PARAGUAY - EXECUTED BUDGET OF OTHER PUBLIC ENTERPRISES, 1980-1990 a/
(Current Guaranies, Millions)

	1980	1981	1982	1983	1984	1985	1986	1987	1988	1989	1990
CURRENT ACCOUNT	425	255	190	443	454	788	931	1256	1537	6271	12584
CURRENT REVENUES	1545	1631	1774	2115	2297	2813	3476	4380	6185	12945	21640
Value Added	1143	1077	1095	1505	1431	2200	2778	3249	4676	11122	21087
Sales of Goods & Services	1700	1782	1853	2176	2412	3505	4207	5613	7325	16016	24701
(Cost of Inputs)	556	706	758	671	981	1305	1429	2364	2649	4894	3614
Other	159	223	393	311	578	202	170	468	756	693	553
Current Transfers	243	332	286	300	288	411	527	663	647	1130	0
Capital Transfers & Loans	0	0	0	0	0	0	0	0	106	0	0
CURRENT EXPENDITURES	1120	1376	1584	1672	1843	2025	2544	3124	4648	6675	9057
Operating Costs	1118	1372	1584	1672	1842	2022	2540	3111	4637	6669	9047
Wages & Salaries	1054	1289	1485	1551	1736	1905	2425	2833	4400	6377	8776
Social Security Contrib.	28	45	49	57	51	33	61	44	75	82	122
Rent	2	5	18	16	16	58	24	29	32	152	138
Interest on Internal Debt	16	9	10	10	24	8	10	122	11	23	34
Interest on External Debt	18	24	21	38	15	18	21	83	119	45	49.0902
Taxes & Transfers	2	4	0	0	1	3	4	13	12	6	10
Taxes	0	0	0	0	0	0	0	4	2	0	0
Transfers	1	4	0	0	1	3	4	9	10	6	10
CAPITAL ACCOUNT	705	788	376	362	405	550	892	1356	6249	970	925
Real Investment	667	728	381	570	376	540	826	1240	6180	840	1065
Change in Inventories	47	44	-5	-51	21	-10	47	-17	65	-11	47
Financial Investment	0	8	0	181	8	21	26	133	4	180	113
Net Loans	0	0	0	0	0	0	0	0	0	0	0
Capital Revenues	9	-8	0	338	0	0	7	0	0	38	300
BALANCE	-280	-532	-186	81	49	237	40	-99	-4712	5300	11659

a/ Includes LATN, FCCAL, ANNP and ANAC.
Source: Luis A. Campos & Ricardo Canese, "El Sector Publico en el Paraguay" and Ministry of Finance.
18-Sep-91

TABLE 5.16: PARAGUAY - TAX REVENUE BY SOURCE, 1970-1990
(In millions of guaranies and as percent of GDP)

	1970	1971	1972	1973	1974	1975	1976	1977	1978	1979	1980	1981	1982	1983	1984	1985	1986	1987	1988	1989	1990
Taxes on Goods and Services	3479	3575	3728	5111	6800	7140	8101	11278	13637	17316	19907	22387	26075	23967	45040	59423	76678	108842	142735	219159	298310
Consumption	2297	2410	2360	2818	3600	3612	4058	5459	6376	7377	9345	8619	10861	9773	24232	32687	43973	60063	76261	99473	142970
General Sales	403	411	442	642	771	832	1047	1593	2126	2878	3619	3743	3690	2916	6014	9194	13613	20390	26923	36433	50489
Selective Sales a/	1894	1999	1918	2276	2729	2779	3011	3866	4248	4499	4828	4876	7291	6857	18218	23493	30460	39663	49428	63040	92481
Stamp Taxes	1182	1165	1368	2293	3300	3528	4043	5819	7261	9939	11462	13768	14194	14194	19163	24720	30295	43975	63067	112659	151398
Other b/															1625	2018	2310	2814	3427	7027	3942
Income Taxes	794	882	945	1141	1724	2394	2404	3207	4937	6493	8837	11104	12653	9561	11848	17742	22970	38193	47227	64699	81390
Capital Taxes															4067	5622	6945	8236	8207	11332	17161
Land Property	399	435	503	572	591	651	755	894	1024	1191	1452	1952	2691	3348	3783	5204	6372	7488	8351	10303	0
Inheritance															274	418	573	748	856	1029	17161
Other Taxes	3031	2910	2671	3168	5109	5301	5435	7601	10580	13719	16541	16606	18371	18690	10805	13481	18619	24620	33826	111408	199372
Import and Export Taxes	2966	2850	2540	3061	4623	4742	4659	6923	9189	12164	13019	12525	10421	8192	10805	13481	18619	24620	33826	95487	160225
Misc. internal taxes a/	65	60	131	107	586	559	776	678	1391	1555	2522	3081	7950	10498	0	0	0	0	0	15919	39147
Total	7702	7802	7847	9991	14224	15485	16895	22979	30178	38720	46636	51048	58590	55567	71760	96268	125012	177891	232995	406596	596233
Taxes on Goods and Services	4.64%	4.27%	3.85%	4.07%	4.05%	3.76%	3.78%	4.28%	4.23%	4.02%	3.53%	3.18%	3.40%	2.93%	4.21%	4.28%	4.18%	4.28%	4.30%	4.76%	4.61%
Consumption	3.07%	2.88%	2.44%	2.25%	2.08%	1.90%	1.90%	2.07%	1.98%	1.71%	1.49%	1.22%	1.48%	1.19%	2.26%	2.35%	2.40%	2.41%	2.30%	2.16%	2.21%
General Sales	0.54%	0.49%	0.45%	0.43%	0.46%	0.44%	0.49%	0.60%	0.65%	0.67%	0.63%	0.53%	0.49%	0.36%	0.56%	0.66%	0.74%	0.82%	0.81%	0.79%	0.78%
Selective Sales a/	2.53%	2.39%	1.96%	1.81%	1.62%	1.46%	1.41%	1.47%	1.32%	1.06%	0.86%	0.69%	0.95%	0.84%	1.70%	1.69%	1.66%	1.59%	1.49%	1.37%	1.43%
Stamp Taxes	1.58%	1.39%	1.41%	1.83%	1.96%	1.85%	1.89%	2.21%	2.25%	2.31%	2.05%	1.94%	1.93%	1.73%	1.79%	1.77%	1.65%	1.76%	1.90%	2.44%	2.34%
Other b/															0.15%	0.14%	0.13%	0.11%	0.10%	0.15%	0.06%
Income Taxes	1.06%	1.05%	0.98%	0.91%	1.03%	1.26%	1.12%	1.22%	1.53%	1.51%	1.58%	1.57%	1.70%	1.17%	1.11%	1.27%	1.25%	1.53%	1.42%	1.40%	1.26%
Capital Taxes															0.38%	0.40%	0.38%	0.33%	0.25%	0.25%	0.27%
Land Property	0.53%	0.52%	0.62%	0.46%	0.35%	0.34%	0.35%	0.34%	0.32%	0.28%	0.26%	0.29%	0.35%	0.41%	0.35%	0.37%	0.35%	0.30%	0.26%	0.22%	0.00%
Inheritance															0.03%	0.03%	0.03%	0.03%	0.03%	0.02%	0.27%
Other Taxes	4.05%	3.47%	2.76%	2.53%	3.04%	2.79%	2.54%	2.88%	3.29%	3.19%	2.77%	2.20%	2.49%	2.28%	1.01%	0.97%	1.01%	0.99%	1.02%	2.42%	3.08%
Import and Export Taxes	3.96%	3.40%	2.62%	2.44%	2.69%	2.49%	2.18%	2.63%	2.85%	2.83%	2.32%	1.77%	1.41%	1.00%	1.01%	0.97%	1.01%	0.99%	1.02%	2.07%	2.47%
Misc. internal taxes a/	0.09%	0.07%	0.14%	0.09%	0.35%	0.29%	0.36%	0.25%	0.43%	0.36%	0.45%	0.43%	1.08%	1.28%	0.00%	0.00%	0.00%	0.00%	0.00%	0.35%	0.60%
Total	10.3%	9.32%	8.10%	7.97%	8.47%	8.13%	7.80%	8.72%	9.36%	8.99%	8.14%	7.20%	7.95%	6.79%	6.70%	6.81%	6.82%	7.13%	7.02%	8.82%	9.21%

a/ Taxes considered as "misc. internal taxes," 1970-1983, included in selective sales taxes 1984-1989.
b/ Includes gambling, tourist and driver's license taxes, 1984-1989.
Source: BCP, Ministry of Finance and World Bank estimates.

27-Sep-91

Table 6.1: PARAGUAY - ACCOUNTS OF THE CONSOLIDATED FINANCIAL SYSTEM, 1980-1990
(Current Guaranies Millions, End of Period)

	1980	1981	1982	1983	1984	1985	1986	1987	1988	1989	1990
I. Consolidated Financial System											
Net Foreign Assets	97289	102910	86701	80914	55206	42584	23165	41034	-13228	247697	584201
Credit	77575	106684	120635	144499	174006	200696	269564	315733	368235	406981	715474
Public Sector	-16023	-4136	1248	19917	23532	34600	42154	35185	9340	-103361	27741
Credit	12522	13225	14267	35055	46704	60956	75607	111692	141630	136929	217378
Deposits	28545	17361	13019	15138	23172	26356	33453	76507	132290	240290	189637
Private Sector	93598	110820	119387	124582	150474	166096	227410	280548	358895	510342	687733
Other Assets, Net	-43803	-58139	-51269	-43069	3541	16984	39132	75237	128677	203952	-134778
Assets = Liabilities	131061	151455	156067	182344	232753	260264	331861	432004	483684	858630	1164897
M1	58058	58419	57391	70453	88539	114780	146504	209300	263062	384340	490655
Currency	31181	31161	33170	38466	48597	62613	84475	119565	149094	216192	300523
Demand Deposits	26877	27258	24221	31987	39942	52167	62029	89735	113968	168148	190132
QM Private Sector	58927	76771	81020	91706	100837	108864	143492	173104	186280	405306	587252
Local Currency	38178	55027	64472	79475	88550	100365	128788	156495	168300	226915	291516
Foreign Currency	20749	21744	16548	12231	12287	8499	14704	16609	17980	178391	295736
LT Foreign Liabilities	14076	16265	17656	20185	43377	36620	41865	49600	34342	68984	86990
Memo:											
Gross Domestic Product	560459	708689	737040	818114	1070444	1393890	1833800	2493601	3319124	4603625	6624031
Net For. Assets ('000 US$)	772134	816749	688106	642174	535059	482467	421782	477621	341968	555841	823971
II. Central Bank											
Net Foreign Assets	94330	98324	81926	78795	52790	43725	17572	30311	-23129	128388	440121
Credit	23238	27454	32473	59949	85728	97208	154234	185313	245848	265815	351788
Public Sector	11507	11681	12110	33248	45351	60146	74972	110371	141180	135913	216884
Banks - Local Currency	11729	15753	20062	26350	39863	36606	78891	74845	104532	129446	134373
Commercial	8599	11124	9280	15053	26078	22094	53298	47886	65202	67294	55040
BNF	3027	3750	10062	10933	12932	11498	22244	22567	30024	45838	63666
Cattle Fund	103	879	720	364	853	3014	3349	4392	9306	16314	15667
Banks - Foreign Currency	2	20	301	351	514	456	371	97	136	456	531
Other Assets, Net	-22451	-29440	-22342	-22920	24549	39041	73277	135405	211091	279858	-71150
Assets = Liabilities	95134	96361	92077	115824	163067	179974	245083	351029	433810	674061	720759
Currency	31181	31161	33170	38466	48597	62613	84475	119565	149094	216192	300523
Cash in Vaults	2531	3208	2961	3382	5106	6950	7585	8744	10181	13715	24745
Official Deposits	23828	11434	7552	8146	14269	16495	21213	34626	64931	183793	119909
Bank Deposits	32813	44444	40964	55510	66423	73936	109053	158576	195036	220829	219561
Local Currency	32796	44421	40944	55510	66423	73936	109053	158576	195036	220829	219561
Foreign Currency	17	23	20	0	0	0	0	0	0	0	0
LT Foreign Liabilities	4781	6114	7430	10320	28672	19980	22757	29518	14568	39532	56021
Memo:											
Net For. Assets ('000 US$)	748650	780352	650209	625361	517004	479234	397506	437314	303716	427910	676303
Assets	766478	810004	686829	684067	669232	578039	475485	526536	337980	449900	700407
Liabilities	17828	29652	36620	58706	152228	98805	77979	89222	34264	21990	24104
Exchange Rate (G$/US$)	126	126	126	126	240	240	320	320	400	1220	1255

Source: Central Bank of Paraguay, Department of Economic Studies.
23-Sep-91

Table 6.1: PARAGUAY - ACCOUNTS OF THE CONSOLIDATED FINANCIAL SYSTEM, 1980-1990
(Current Guaranies Millions, End of Period)

III. Commercial Banks

	1980	1981	1982	1983	1984	1985	1986	1987	1988	1989	1990
Net Foreign Assets	2987	6048	5090	1468	2748	-572	6163	11266	10449	119327	144751
Reserves	29968	41412	39195	53396	63248	70723	94497	127272	161696	204323	207679
Cash in Vault	1677	2347	2202	2367	3827	5017	5439	5675	7206	9341	16742
Deposits with BCP	28291	39065	36993	51029	59421	65706	89058	121597	154490	194982	190937
Local Currency	28291	39065	36993	51029	59421	65706	89058	121597	154490	194982	190937
Foreign Currency	0	0	0	0	0	0	0	0	0	0	0
Credit	73746	85398	90355	92541	109237	117533	160670	203510	259920	374869	525760
Public Sector	1015	1544	2157	1807	1353	810	635	1321	450	1016	494
Private Sector	72731	83854	88198	90734	107884	116723	160035	202189	259470	373853	525266
Other Assets, Net	-15633	-20233	-21619	-10753	-11353	-9087	-12963	-25794	-40121	-36146	-18716
Assets = Liabilities	91068	112625	113021	136652	163880	178597	248367	316254	391944	662373	859474
Demand Deposits	24165	25220	21736	29808	37778	48286	55559	95217	135115	187477	199364
Public Sector	730	1052	709	1912	928	1056	1184	14031	34167	31942	21807
Private Sector	23435	24168	21027	27896	36850	47230	54375	81186	100948	155535	177557
QM	58296	75785	81108	91179	98637	107120	139225	173151	191627	407602	605070
Public Sector	2294	3254	3650	3863	3762	5434	6188	16904	15385	14177	25830
Private Sector	56002	72531	77458	87316	94875	101686	133037	156247	176242	393425	579240
Local Currency	35253	50787	60910	75085	82588	93187	118333	139638	158262	215034	283504
Foreign Currency	20749	21744	16548	12231	12287	8499	14704	16609	17980	178391	295736
Credit from BCP	8601	11144	9581	15255	26499	22513	53583	47886	65202	67294	55040
Local Currency	8599	11124	9280	15053	26078	22094	53298	47886	65202	67294	55040
Foreign Currency	2	20	301	202	421	419	285	0	0	0	0
LT Foreign Liabilities	6	476	596	410	966	678	0	0	0	0	0
Memo:											
Net For. Assets ('000 US$)	23706	48000	40397	11650	16984	3150	24198	40144	38102	127346	147604
Assets	52762	93667	78135	82588	70792	57413	54045	53753	58965	137347	163797
Liabilities	29056	45667	37738	70938	53808	54263	29847	13609	20863	10001	16193
Exchange Rate (G$/US$)	126	126	126	126	240	240	320	320	400	1220	1255

IV. Banco Nacional de Fomento

	1980	1981	1982	1983	1984	1985	1986	1987	1988	1989	1990
Net Foreign Assets	-28	-1462	-315	651	-332	-569	-570	-543	-548	-18	-671
Reserves	5376	6240	4730	5496	8281	10163	22141	40048	43521	30221	36627
Cash in Vault	854	861	759	1015	1279	1933	2146	3069	2975	4374	8003
Deposits with BCP	4522	5379	3971	4481	7002	8230	19995	36979	40546	25847	28624
Local Currency	4505	5356	3951	4481	7002	8230	19995	36979	40546	25847	28624
Foreign Currency	17	23	20	0	0	0	0	0	0	0	0
Private Sector Credit	20764	26087	30469	33484	41737	46359	64026	73967	90119	120175	146800
Other Assets, Net	-5736	-8489	-7328	-9396	-9655	-12970	-21182	-34374	-42293	-39760	-44912
Assets = Liabilities	20376	22376	27556	30235	40031	42983	64415	79098	90799	110618	137844
Demand Deposits	4400	4056	3968	5111	6979	7961	11884	18515	29463	21477	30422
Public Sector	958	966	774	1020	3887	3024	4230	9966	16443	8864	17847
Private Sector	3442	3090	3194	4091	3092	4937	7654	8549	13020	12613	12575
QM	3660	4895	3896	4587	6288	7525	11093	17837	11402	13395	12256
Public Sector	735	655	334	197	326	347	638	980	1364	1514	4244
Private Sector	2925	4240	3562	4390	5962	7178	10455	16857	10038	11881	8012
Credit from BCP	3027	3750	10062	11082	13025	11535	22330	22664	30160	46294	64197
Local Currency	3027	3750	10062	10933	12932	11498	22244	22567	30024	45838	63666
Foreign Currency	0	0	0	149	93	37	86	97	136	456	531
LT Foreign Liabilities	9289	9675	9630	9455	13739	15962	19108	20082	19774	29452	30969
Memo:											
Net For. Assets ('000 US$)	-222	-11603	-2500	5163	1071	83	78	163	150	585	64
Assets	984	540	2540	5163	1071	83	78	163	150	585	64
Liabilities	1206	12143	5040	0	0	0	0	0	0	0	0
Exchange Rate (G$/US$)	126	126	126	126	240	240	320	320	400	1220	1255

Source: Central Bank of Paraguay, Department of Economic Studies.
23-Sep-91

Table 7.1: PARAGUAY : CONSUMER PRICE INDEX, 1972-1991
(1980 = 100)

page 1 of 2

	Index				General Index	Growth Rate				General Index
	Food	Housing	Clothing	Others		Food	Housing	Clothing	Others	
1972	31.2	38.5	36.6	34.1	33.5					
1973	38.0	41.2	38.6	34.4	37.7	21.7	6.9	5.6	0.8	12.8
1974	47.4	51.5	46.6	44.3	47.2	24.8	25.0	20.8	28.7	25.2
1975	49.6	56.2	52.7	47.6	50.4	4.6	9.2	13.0	7.5	6.7
1976	51.7	58.0	55.8	50.2	52.7	4.2	3.1	5.9	5.5	4.5
1977	57.5	62.6	60.0	53.6	57.6	11.3	8.0	7.5	6.7	9.4
1978	64.9	66.6	67.0	57.7	63.7	13.0	6.4	11.7	7.6	10.6
1979	84.1	81.5	82.5	75.7	81.7	29.5	22.4	23.2	31.3	28.2
1980	100.0	100.0	100.0	100.0	100.0	18.9	22.8	21.2	32.1	22.4
1981	110.4	120.2	111.6	113.4	114.0	10.4	20.2	11.6	13.4	14.0
1982	114.3	130.3	118.3	124.1	121.7	3.5	8.4	6.0	9.4	6.8
1983	133.9	135.0	142.9	145.5	138.1	17.1	3.6	20.8	17.2	13.5
1984	172.6	144.5	181.0	174.6	166.1	28.9	7.0	26.7	20.0	20.3
January	147.2	136.9	159.3	156.4	147.9	3.2	0.3	0.6	1.0	1.6
February	151.3	136.8	161.2	157.4	149.8	2.8	-0.1	1.2	0.6	1.3
March	157.9	137.5	162.7	159.6	153.1	4.4	0.5	0.9	1.4	2.2
April	158.9	137.8	165.1	161.5	154.3	0.6	0.2	1.5	1.2	0.8
May	158.0	139.2	169.1	164.4	155.5	-0.6	1.0	2.4	1.8	0.8
June	169.6	140.9	175.2	168.1	161.7	7.3	1.2	3.6	2.3	4.0
July	169.9	142.2	177.8	176.3	164.6	0.2	0.9	1.5	4.9	1.8
August	178.5	144.4	187.9	183.8	171.3	5.1	1.5	5.7	4.3	4.1
September	182.2	149.2	197.7	188.0	176.0	2.1	3.3	5.2	2.3	2.7
October	192.0	154.0	202.1	192.1	182.4	5.4	3.2	2.2	2.2	3.6
November	202.4	157.0	204.5	193.2	187.6	5.4	1.9	1.2	0.6	2.9
December	203.4	158.2	209.3	194.2	189.0	0.5	0.8	2.3	0.5	0.7
1985	220.1	169.7	240.4	220.3	208.0	27.5	17.4	32.8	26.2	25.2
January	200.9	159.4	213.0	193.4	188.5	-1.2	0.8	1.8	-0.4	-0.3
February	200.0	160.4	214.4	196.5	189.4	-0.4	0.6	0.7	1.6	0.5
March	203.8	161.4	214.7	199.5	191.9	1.9	0.6	0.1	1.5	1.3
April	202.6	161.7	216.0	201.9	192.3	-0.6	0.2	0.6	1.2	0.2
May	207.1	162.8	217.7	203.9	194.9	2.2	0.7	0.8	1.0	1.4
June	206.0	165.5	222.9	209.2	197.1	-0.5	1.7	2.4	2.6	1.1
July	202.8	167.0	228.0	214.8	198.3	-1.6	0.9	2.3	2.7	0.6
August	233.0	177.8	252.9	229.5	218.5	14.9	6.5	10.9	6.8	10.2
September	239.0	180.8	267.1	244.4	226.8	2.6	1.7	5.6	6.5	3.8
October	249.3	179.7	275.4	251.1	232.8	4.3	-0.6	3.1	2.7	2.6
November	248.1	180.7	280.0	249.5	232.6	-0.5	0.6	1.7	-0.6	-0.1
December	248.8	178.8	282.4	249.6	232.6	0.3	-1.1	0.9	0.0	0.0
1986	316.0	193.1	319.5	284.7	274.0	43.6	13.8	32.9	29.2	31.8
January	273.8	182.4	283.2	259.7	245.6	10.0	2.0	0.3	4.0	5.6
February	303.4	187.0	286.7	268.3	260.4	10.8	2.5	1.2	3.3	6.0
March	315.3	190.1	291.6	277.3	268.4	3.9	1.7	1.7	3.4	3.1
April	315.4	191.3	306.1	278.3	270.4	0.0	0.6	5.0	0.4	0.7
May	313.2	193.0	318.1	282.5	272.2	-0.7	0.9	3.9	1.5	0.7
June	316.4	193.6	324.2	286.2	275.1	1.0	0.3	1.9	1.3	1.1
July	308.9	193.9	330.9	286.4	273.1	-2.4	0.2	2.1	0.1	-0.7
August	314.7	193.8	334.3	292.4	277.1	1.9	-0.1	1.0	2.1	1.5
September	323.9	196.4	337.1	294.6	282.0	2.9	1.3	0.8	0.8	1.8
October	339.2	197.9	339.3	295.6	288.6	4.7	0.8	0.7	0.3	2.3
November	332.3	198.4	340.2	296.7	286.5	-2.0	0.3	0.3	0.4	-0.7
December	335.7	199.7	342.2	298.1	288.7	1.0	0.7	0.6	0.5	0.8

Source: Central Bank of Paraguay.
23-Sep-91

Table 7.1: PARAGUAY : CONSUMER PRICE INDEX, 1972-1991
(1980 = 100)

page 2 of 2

	Index					Growth Rate				
	Food	Housing	Clothing	Others	General Index	Food	Housing	Clothing	Others	General Index
1987	391.2	229.7	379.6	347.1	333.8	23.8	18.9	18.8	21.9	21.8
January	353.3	212.7	345.2	316.2	303.8	5.2	6.5	0.9	6.1	5.2
February	372.4	217.3	347.4	321.8	313.9	5.4	2.2	0.6	1.8	3.3
March	350.7	223.4	353.1	330.8	310.4	-5.8	2.8	1.6	2.8	-1.1
April	347.4	225.6	365.2	334.9	311.9	-0.9	1.0	3.4	1.2	0.5
May	365.3	225.7	370.9	340.9	320.7	5.2	0.0	1.6	1.8	2.8
June	373.0	229.2	377.6	348.5	327.1	2.1	1.6	1.8	2.2	2.0
July	363.0	232.2	386.2	350.3	325.5	-2.7	1.3	2.3	0.5	-0.5
August	378.8	233.8	388.9	352.2	332.5	4.4	0.7	0.7	0.5	2.2
September	400.4	235.1	392.4	357.0	342.4	5.7	0.6	0.9	1.4	3.0
October	431.1	237.0	399.9	362.9	356.5	7.7	0.8	1.9	1.7	4.1
November	480.8	240.3	410.0	372.0	379.1	11.5	1.4	2.5	2.5	6.3
December	477.8	243.6	417.8	377.9	381.2	-0.6	1.4	1.9	1.6	0.6
1988	484.2	275.2	462.4	432.1	410.6	23.8	19.8	21.8	24.5	23.0
January	475.1	246.9	422.4	392.5	385.3	-0.6	1.4	1.1	3.9	1.1
February	505.8	252.2	429.5	405.0	402.1	6.5	2.1	1.7	3.2	4.4
March	441.2	254.8	435.4	415.9	382.3	-12.8	1.0	1.4	2.7	-4.9
April	444.5	259.9	442.6	422.8	387.4	0.7	2.0	1.7	1.7	1.3
May	446.7	266.0	450.1	427.4	391.8	0.5	2.3	1.7	1.1	1.1
June	451.7	268.8	462.0	429.1	396.0	1.1	1.1	2.6	0.4	1.1
July	472.2	279.3	468.8	432.7	408.0	4.5	3.9	1.5	0.8	3.0
August	480.2	285.0	473.8	437.3	414.2	1.7	2.0	1.1	1.1	1.5
September	516.6	290.2	481.4	443.4	431.4	7.6	1.8	1.6	1.4	4.2
October	530.4	295.3	487.3	450.1	440.2	2.7	1.8	1.2	1.5	2.0
November	521.7	300.9	494.7	463.4	442.7	-1.6	1.9	1.5	3.0	0.6
December	524.6	303.6	500.9	466.1	445.8	0.6	0.9	1.3	0.6	0.7
1989	582.9	368.2	576.0	561.9	517.3	20.4	33.8	24.6	30.0	26.0
January	513.1	312.7	507.6	486.9	449.9	-2.2	3.0	1.3	4.2	0.9
February	530.0	320.5	515.1	499.7	462.6	3.3	2.5	1.5	2.8	2.8
March	536.5	325.4	534.8	511.9	471.3	1.2	1.5	3.8	2.4	1.9
April	529.8	330.5	541.4	516.9	472.2	-1.2	1.6	1.2	1.0	0.2
May	538.9	338.6	548.9	534.8	483.2	1.7	2.5	1.4	3.5	2.3
June	550.0	381.8	565.8	564.2	508.5	2.1	12.8	3.1	5.5	5.2
July	580.3	389.1	577.4	586.5	528.7	5.5	1.9	2.1	4.0	4.0
August	620.5	395.4	587.1	596.0	548.7	6.9	1.6	1.7	1.6	3.8
September	659.2	401.9	614.6	604.8	569.5	6.2	1.6	4.7	1.5	3.8
October	643.9	403.1	629.7	608.1	566.7	-2.3	0.3	2.5	0.5	-0.5
November	650.4	407.7	640.0	614.5	572.7	1.0	1.1	1.6	1.1	1.1
December	642.2	412.1	650.0	619.0	573.0	-1.3	1.1	1.6	0.7	0.1
1990	853.4	494.3	763.8	735.4	714.8	46.4	34.2	32.6	30.9	38.2
January	664.1	431.0	665.9	639.4	593.1	3.4	4.6	2.4	3.3	3.5
February	698.5	443.4	680.4	663.2	616.9	5.2	2.9	2.2	3.7	4.0
March	739.1	453.8	695.5	670.0	637.9	5.8	2.3	2.2	1.0	3.4
April	771.4	467.5	716.9	690.5	661.0	4.4	3.0	3.1	3.1	3.6
May	798.3	478.5	730.7	701.8	678.2	3.5	2.4	1.9	1.6	2.6
June	848.4	484.7	742.0	709.1	701.3	6.3	1.3	1.5	1.0	3.4
July	885.9	488.5	755.0	717.7	719.7	4.4	0.8	1.8	1.2	2.6
August	933.4	498.0	774.1	731.4	745.2	5.4	1.9	2.5	1.9	3.5
September	951.3	509.4	805.2	786.1	772.2	1.9	2.3	4.0	7.5	3.6
October	983.0	551.1	846.7	823.6	809.1	3.3	8.2	5.2	4.8	4.8
November	984.5	559.4	869.1	834.8	816.9	0.2	1.5	2.6	1.4	1.0
December	983.4	566.3	883.7	856.7	825.5	-0.1	1.2	1.7	2.6	1.1
1991	6432.8	596.7	928.3	909.9	864.9	868.7	38.5	39.4	42.3	45.8
January	1013.7	572.5	895.4	870.5	843.0	3.1	1.1	1.3	1.6	2.1
February	1028.2	584.6	915.8	889.0	858.5	1.4	2.1	2.3	2.1	1.8
March	1036	601.1	922.7	918.9	874.4	0.8	2.8	0.8	3.4	1.9
April	1010.8	605.3	948.1	927.6	870.8	-2.4	0.7	2.8	0.9	-0.4
May	1003.8	620.1	959.3	943.4	877.6	-0.7	2.4	1.2	1.7	0.8

Source: Central Bank of Paraguay.
23-Sep-91

Table 7.1: PARAGUAY : CONSUMER PRICE INDEX, 1972-1991
(1980 = 100)

	Index					Growth Rate				
	Food	Housing	Clothing	Others	General Index	Food	Housing	Clothing	Others	General Index
1987	391.2	229.7	379.6	347.1	333.8	23.8	18.9	18.8	21.9	21.8
January	353.3	212.7	345.2	316.2	303.8	5.2	6.5	0.9	6.1	5.2
February	372.4	217.3	347.4	321.8	313.9	5.4	2.2	0.6	1.8	3.3
March	350.7	223.4	353.1	330.8	310.4	-5.8	2.8	1.6	2.8	-1.1
April	347.4	225.6	365.2	334.9	311.9	-0.9	1.0	3.4	1.2	0.5
May	365.3	225.7	370.9	340.9	320.7	5.2	0.0	1.6	1.8	2.8
June	373.0	229.2	377.6	348.5	327.1	2.1	1.6	1.8	2.2	2.0
July	363.0	232.2	386.2	350.3	325.5	-2.7	1.3	2.3	0.5	-0.5
August	378.8	233.8	388.9	352.2	332.5	4.4	0.7	0.7	0.5	2.2
September	400.4	235.1	392.4	357.0	342.4	5.7	0.6	0.9	1.4	3.0
October	431.1	237.0	399.9	362.9	356.5	7.7	0.8	1.9	1.7	4.1
November	480.8	240.3	410.0	372.0	379.1	11.5	1.4	2.5	2.5	6.3
December	477.8	243.6	417.8	377.9	381.2	-0.6	1.4	1.9	1.6	0.6
1988	484.2	275.2	462.4	432.1	410.6	23.8	19.8	21.8	24.5	23.0
January	475.1	246.9	422.4	392.5	385.3	-0.6	1.4	1.1	3.9	1.1
February	505.8	252.2	429.5	405.0	402.1	6.5	2.1	1.7	3.2	4.4
March	441.2	254.8	435.4	415.9	382.3	-12.8	1.0	1.4	2.7	-4.9
April	444.5	259.9	442.6	422.8	387.4	0.7	2.0	1.7	1.7	1.3
May	446.7	266.0	450.1	427.4	391.8	0.5	2.3	1.7	1.1	1.1
June	451.7	268.8	462.0	429.1	396.0	1.1	1.1	2.6	0.4	1.1
July	472.2	279.3	468.8	432.7	408.0	4.5	3.9	1.5	0.8	3.0
August	480.2	285.0	473.8	437.3	414.2	1.7	2.0	1.1	1.1	1.5
September	516.6	290.2	481.4	443.4	431.4	7.6	1.8	1.6	1.4	4.2
October	530.4	295.3	487.3	450.1	440.2	2.7	1.8	1.2	1.5	2.0
November	521.7	300.9	494.7	463.4	442.7	-1.6	1.9	1.5	3.0	0.6
December	524.6	303.6	500.9	466.1	445.8	0.6	0.9	1.3	0.6	0.7
1989	582.9	368.2	576.0	561.9	517.3	20.4	33.8	24.6	30.0	26.0
January	513.1	312.7	507.6	485.9	449.9	-2.2	3.0	1.3	4.2	0.9
February	530.0	320.5	515.1	499.7	462.6	3.3	2.5	1.5	2.8	2.8
March	536.5	325.4	534.8	511.9	471.3	1.2	1.5	3.8	2.4	1.9
April	529.8	330.5	541.4	516.9	472.7	-1.2	1.6	1.2	1.0	0.2
May	538.9	338.6	548.9	534.8	483.2	1.7	2.5	1.4	3.5	2.3
June	550.0	381.8	565.8	564.2	508.5	2.1	12.8	3.1	5.5	5.2
July	580.3	389.1	577.4	586.5	528.7	5.5	1.9	2.1	4.0	4.0
August	620.5	395.4	587.1	596.0	548.7	6.9	1.6	1.7	1.6	3.8
September	659.2	401.9	614.6	604.8	569.5	6.2	1.6	4.7	1.5	3.8
October	643.9	403.1	629.7	608.1	566.7	-2.3	0.3	2.5	0.5	-0.5
November	650.4	407.7	640.0	614.5	572.7	1.0	1.1	1.6	1.1	1.1
December	642.2	412.1	650.0	619.0	573.0	-1.3	1.1	1.6	0.7	0.1
1990	853.4	494.3	763.8	735.4	714.8	46.4	34.2	32.6	30.9	38.2
January	664.1	431.0	665.9	639.4	593.1	3.4	4.6	2.4	3.3	3.5
February	698.5	443.4	680.4	663.2	616.9	5.2	2.9	2.2	3.7	4.0
March	739.1	453.8	695.5	670.0	637.9	5.8	2.3	2.2	1.0	3.4
April	771.4	467.5	716.9	690.5	661.0	4.4	3.0	3.1	3.1	3.6
May	798.3	478.5	730.7	701.8	678.2	3.5	2.4	1.9	1.6	2.6
June	848.4	484.7	742.0	709.1	701.3	6.3	1.3	1.5	1.0	3.4
July	885.9	488.5	755.0	717.7	719.7	4.4	0.8	1.8	1.2	2.6
August	933.4	498.0	774.1	731.4	745.2	5.4	1.9	2.5	1.9	3.5
September	951.3	509.4	805.2	786.1	772.2	1.9	2.3	4.0	7.5	3.6
October	983.0	551.1	846.7	823.6	809.1	3.3	8.2	5.2	4.8	4.8
November	984.5	559.4	869.1	834.8	816.9	0.2	1.5	2.6	1.4	1.0
December	983.4	566.3	883.7	856.7	825.5	-0.1	1.2	1.7	2.6	1.1
1991	6432.8	596.7	928.3	909.9	864.9	868.7	38.5	39.4	42.3	45.8
January	1013.7	572.5	895.4	870.5	843.0	3.1	1.1	1.3	1.6	2.1
February	1028.2	584.6	915.8	889.0	858.5	1.4	2.1	2.3	2.1	1.8
March	1036	601.1	922.7	918.9	874.4	0.8	2.8	0.8	3.4	1.9
April	1010.8	605.3	948.1	927.6	870.8	-2.4	0.7	2.8	0.9	-0.4
May	1003.8	620.1	959.3	943.4	877.6	-0.7	2.4	1.2	1.7	0.8

Source: Central Bank of Paraguay.
23-Sep-91

Table 7.2: PARAGUAY - WHOLESALE PRICE INDEX 1972-1990
(1980 = 100)

continues...

	General Index	Agriculture and Hunting	Forestry	Food Beverages & Tobacco	Textiles and Leather	Wood Products	Paper and Printing	Chemicals	Nonmetals Excluding Petroleum	Basic Metals	Machinery and Equipment	Other anufacture	Electricity and Steam
1972	28.8	19.7	40.1	44.8	34.2	31.6	29.7	30.2	30.4	40.9	47.4	23.1	77.5
1973	39.7	34.6	50.8	51.7	40.7	39.4	31.1	36.3	33.9	51.9	50.1	25.5	77.5
1974	51.8	41.3	69.2	71.8	50.5	64.7	40.2	64.5	46.3	89.4	64.3	47.2	77.5
1975	59.6	52.2	66.6	73.0	50.9	67.3	50.1	65.4	49.8	99.3	80.8	72.7	77.5
1976	60.3	53.5	61.8	74.2	53.1	61.2	51.3	65.1	50.0	84.5	83.1	73.0	92.2
1977	65.1	57.9	72.1	75.4	58.9	67.5	52.5	83.2	54.7	79.5	85.5	72.8	110.7
1978	73.5	70.4	76.5	78.4	66.8	73.3	60.8	81.6	64.0	77.4	86.0	78.0	121.7
1979	92.8	97.4	84.6	87.7	84.5	82.8	72.4	85.9	86.0	92.0	91.9	90.1	119.0
1980	100.0	100.0	100.0	100.0	100.0	100.0	100.0	100.0	100.0	100.0	100.0	100.0	100.0
1981	112.2	113.7	109.7	109.3	112.5	100.6	119.3	117.3	108.3	108.4	108.7	107.8	100.0
1982	116.1	117.5	106.5	112.9	118.2	99.5	126.5	118.4	105.2	133.9	117.1	122.6	100.0
1983	141.6	149.4	114.9	125.9	138.3	104.0	150.0	129.9	120.4	170.3	145.1	225.9	100.0
1984	181.6	177.2	179.9	165.8	213.8	147.1	214.7	200.1	162.9	294.3	202.2	293.3	106.2
January	150.2	149.6	126.1	142.0	162.1	112.1	191.2	143.7	133.9	215.8	170.9	269.1	100.0
February	151.2	151.0	126.1	141.6	164.9	112.1	194.8	143.9	134.0	221.0	171.0	269.1	100.0
March	160.3	166.3	126.3	143.8	170.6	112.1	203.0	152.2	143.7	228.2	167.4	269.1	100.0
April	158.3	159.9	126.3	148.1	171.0	112.1	203.0	153.6	146.0	237.8	167.5	269.1	100.0
May	168.9	163.8	166.8	154.9	202.4	146.0	203.0	183.4	149.4	243.6	193.2	282.1	100.0
June	171.0	161.6	166.9	160.0	219.0	145.0	217.1	186.0	160.9	337.8	196.6	282.1	100.0
July	186.8	173.0	208.1	171.6	236.0	169.2	217.1	243.3	174.8	341.3	213.4	313.0	100.0
August	183.9	165.4	208.6	175.4	239.0	169.2	217.1	250.0	174.3	341.3	216.6	313.0	105.0
September	191.0	176.0	225.5	177.4	236.9	169.2	218.0	243.9	180.2	341.3	227.1	313.0	109.9
October	214.8	217.2	225.6	179.6	251.7	170.6	218.0	232.9	186.4	341.3	231.5	313.0	114.9
November	220.4	220.7	225.9	195.0	256.1	174.2	247.3	234.0	186.4	341.3	236.0	313.0	119.9
December	222.6	223.0	226.1	200.4	256.1	174.2	247.3	233.8	184.9	341.3	236.0	313.0	125.0
1985	224.2	200.0	244.6	231.9	304.6	180.0	262.6	247.1	208.5	495.3	264.9	412.7	130.0
January	234.8	241.5	231.2	208.6	252.4	164.2	246.8	241.4	184.0	435.4	235.2	274.2	130.0
February	217.0	206.1	231.2	217.1	253.4	164.2	248.9	251.9	185.3	435.4	228.5	274.2	130.0
March	207.4	187.5	229.5	219.5	261.0	167.9	249.5	246.3	184.5	433.8	229.3	282.0	130.0
April	211.8	195.0	229.5	220.4	261.3	167.9	248.7	246.4	185.4	433.8	231.2	282.0	130.0
May	205.8	178.3	230.0	225.0	283.8	171.5	251.5	241.2	196.8	436.6	241.8	319.2	130.0
June	201.1	168.4	230.0	226.8	289.4	171.5	251.5	243.4	197.3	436.6	242.7	319.2	130.0
July	209.4	168.9	269.6	233.3	292.6	186.8	262.6	246.3	229.7	514.0	279.1	634.9	130.0
August	218.3	173.2	269.6	243.3	331.4	219.0	266.0	249.9	238.0	530.5	300.3	590.1	130.0
September	238.4	205.1	268.5	248.9	344.7	189.0	266.0	250.8	235.7	586.9	304.3	590.1	130.0
October	246.4	219.5	269.0	256.3	344.7	189.0	257.7	251.1	236.4	581.5	300.3	496.3	130.0
November	240.9	212.4	258.4	240.1	370.4	200.6	301.5	249.1	217.2	559.5	293.5	495.5	130.0
December	258.8	244.1	258.4	243.5	370.4	189.0	300.5	249.0	211.5	559.5	292.6	495.5	130.0
1986	325.4	327.9	343.0	287.8	442.0	263.8	373.1	262.6	270.8	639.2	330.9	477.9	131.4
January	295.9	305.1	256.6	250.7	383.7	194.5	329.5	261.6	219.6	558.6	302.1	480.2	130.0
February	311.8	328.4	267.5	267.8	398.7	198.0	328.8	259.7	247.6	569.5	307.8	480.2	130.0
March	320.9	332.8	269.2	278.9	415.3	206.3	380.9	258.9	260.3	568.6	321.1	480.2	130.0
April	309.1	310.6	287.5	274.3	422.0	213.4	380.9	258.8	265.1	568.6	318.5	480.2	130.0
May	319.4	326.0	288.9	280.7	437.2	221.8	380.9	259.4	270.8	568.6	315.0	480.2	130.0
June	316.2	316.2	289.0	291.1	443.4	221.8	380.9	259.4	275.9	568.6	313.4	480.2	130.0
July	320.3	315.3	378.3	294.2	457.7	278.9	381.5	260.6	286.8	511.2	308.5	480.2	130.0
August	330.1	325.2	397.0	291.7	468.9	300.3	381.5	259.7	285.3	511.2	357.7	480.2	130.0
September	355.5	366.0	400.5	308.3	469.8	311.1	382.1	259.8	285.8	511.2	355.9	480.2	130.0
October	355.6	363.7	418.3	311.1	469.8	328.9	382.1	260.5	284.5	511.2	351.6	480.2	130.0
November	339.0	331.4	436.8	303.9	469.0	345.1	384.1	263.2	284.6	511.2	359.8	466.3	130.0
December	331.6	314.6	436.8	312.4	469.0	345.1	384.1	289.2	283.2	511.2	359.8	466.3	147.6

Source: Central Bank of Paraguay.
16-Oct-91

Table 7.2: PARAGUAY - WHOLESALE PRICE INDEX 1972-1990
(1980 = 100)

continued

	General Index	Agriculture and Hunting	Forestry	Food Beverages & Tobacco	Textiles and Leather	Wood Products	Paper and Printing	Chemicals	Nonmetals Excluding Petroleum	Basic Metals	Machinery and Equipment	Other anufacture	Electricity and Steam
1987	364.0	324.5	579.2	354.9	566.5	444.1	454.7	324.5	319.9	534.7	416.3	509.6	182.1
January	336.8	311.7	487.3	309.3	491.8	413.0	419.6	311.5	294.2	526.3	373.2	488.9	168.0
February	341.7	312.6	487.3	318.5	521.5	413.0	419.6	311.9	302.7	545.1	384.7	488.9	183.3
March	342.2	309.3	489.1	319.1	529.5	413.0	447.1	314.4	304.6	545.1	393.9	561.7	183.4
April	339.5	302.5	489.5	322.6	529.5	416.6	447.1	314.4	306.8	545.1	398.7	561.7	183.4
May	342.6	289.2	589.1	352.6	551.2	438.6	447.1	316.9	314.2	554.6	408.4	523.5	183.4
June	349.7	297.5	589.1	359.4	571.9	438.6	447.1	318.1	317.8	554.6	409.6	523.5	183.4
July	354.9	300.9	629.3	361.5	576.1	456.4	447.1	332.2	328.8	504.8	422.1	497.5	183.4
August	354.4	299.9	635.7	360.4	576.9	456.4	447.3	333.5	328.1	505.4	422.7	497.5	183.4
September	380.4	337.7	637.9	373.9	601.0	467.5	449.6	335.8	333.4	519.7	433.0	497.5	183.4
October	406.8	378.2	637.9	389.3	614.1	467.5	451.9	336.7	332.0	533.0	440.3	497.5	183.4
November	421.7	400.0	639.3	394.2	617.0	474.6	517.1	334.1	338.1	543.2	455.1	488.9	183.4
December	397.0	354.3	639.3	398.2	617.0	474.6	516.1	334.3	338.4	540.0	453.6	488.9	183.4
1988	462.4	439.0	644.3	439.2	653.1	485.6	563.8	368.4	429.7	574.0	524.7	530.9	209.8
January	409.4	364.7	639.2	406.5	624.1	459.2	525.4	360.0	362.0	547.6	495.3	552.6	183.4
February	422.9	385.3	640.5	414.7	626.9	459.2	525.6	361.0	371.6	547.6	500.0	552.6	183.4
March	437.4	407.7	641.2	423.2	635.2	463.2	524.8	366.5	378.4	555.0	498.1	519.5	183.4
April	427.3	382.4	641.2	428.4	650.8	468.5	534.8	367.2	397.1	555.0	523.0	519.5	183.4
May	424.8	375.0	642.7	431.4	651.6	468.5	534.8	369.8	429.7	555.0	526.5	519.5	183.4
June	404.2	334.1	643.2	437.2	658.0	468.5	535.0	369.1	440.1	555.0	529.8	552.0	200.9
July	426.1	376.8	644.1	432.0	663.2	495.3	547.4	375.6	459.9	556.8	524.4	519.5	210.8
August	458.7	430.4	644.1	433.7	663.2	495.3	590.1	377.7	459.9	556.8	524.7	529.0	220.4
September	507.4	512.6	643.5	446.7	664.9	496.2	590.1	378.3	462.1	609.0	532.1	519.5	230.0
October	522.0	535.0	650.1	459.5	666.3	510.9	591.3	363.6	456.8	609.0	536.3	529.0	239.6
November	576.2	623.5	651.0	478.4	666.3	521.1	632.9	365.7	485.6	619.1	552.2	529.0	249.2
December	530.8	540.5	651.0	478.8	666.5	521.1	633.5	366.7	473.3	622.7	554.0	529.0	249.2
1989	563.9	518.2	715.0	560.7	778.5	616.8	679.3	499.4	587.0	804.1	676.5	683.2	364.7
January	534.0	528.9	654.6	501.6	676.9	524.8	642.2	371.2	491.8	682.1	596.0	671.4	274.1
February	530.2	518.6	654.6	503.1	677.5	524.8	641.5	375.1	499.5	682.1	613.8	671.4	299.1
March	543.1	536.7	654.7	502.2	700.0	558.0	637.7	382.2	498.1	696.4	618.0	671.4	323.9
April	539.3	529.0	655.8	501.7	700.7	558.0	638.3	384.2	501.7	696.4	622.3	671.4	323.9
May	543.9	507.3	682.8	518.9	725.7	606.0	647.8	485.8	574.5	710.6	620.7	671.4	323.9
June	531.1	466.9	697.9	545.5	766.6	626.0	657.9	522.7	629.9	750.3	651.4	606.9	401.2
July	519.3	430.7	736.1	554.5	782.8	631.8	682.7	524.4	647.0	786.0	699.2	622.8	405.1
August	553.1	473.8	736.1	591.2	783.6	631.8	707.5	534.3	647.0	863.7	733.0	646.8	405.1
September	574.3	492.9	736.1	594.0	869.4	656.4	710.3	607.2	638.8	928.0	733.4	672.3	405.1
October	596.6	506.5	773.6	661.2	884.8	678.5	705.3	602.2	638.8	959.6	735.2	704.6	405.1
November	632.0	579.5	774.3	631.8	885.8	685.9	732.8	602.1	637.8	964.8	733.5	693.5	405.1
December	669.5	648.8	823.8	602.6	888.0	719.1	747.7	600.7	638.8	969.4	762.5	688.2	405.1
1990	874.5	916.4	974.4									814.7	464.4
January	699.0	695.1	801.0									699.1	405.1
February	718.1	696.8	817.2									741.2	443.7
March	762.7	763.6	871.2									756.2	445.7
April	787.0	790.7	908.5									776.1	445.7
May	766.7	751.7	975.8									773.1	445.7
June	823.8	831.3	1021.3									802.9	445.7
July	884.8	929.7	1020.9									818.9	445.7
August	888.0	934.0	1017.3									821.1	445.7
September	930.3	982.0	1093.9									853.7	445.7
October	976.2	1031.0	1095.7									898.3	530.5
November	1138.3	1314.5	1057.4									913.1	534.8
December	1119.5	1276.4	1012.4									922.1	539.0
1991													
January	1141.9	1317.0	1027.4									921.4	538.8
February	1190.7	1389.1	1030.8									942.7	538.8
March	1216.3	1437.3	1027.7									940.4	538.8
April	1214.1	1425.9	1049.6									948.9	538.8
May	1161.1	1279.4	1052.6									1016.7	538.8

Source: Central Bank of Paraguay.

Table 7.3: PARAGUAY - WHOLESALE PRICE INDEX 1972-1991
(Growth Rates)

— continues...

	General Index	Agriculture and Hunting	Forestry	Food Beverages & Tobacco	Textiles and Leather	Wood Products	Paper and Printing	Chemicals	Nonmetals Excluding Petroleum	Basic Metals	Machinery and Equipment	Other anufacture	Electricity and Steam
1973	38.0	75.0	26.8	15.3	18.9	24.7	4.5	20.1	11.8	26.7	5.8	10.4	0.0
1974	30.5	19.4	36.0	38.9	24.1	64.2	29.3	77.7	36.4	72.3	28.3	84.9	0.0
1975	15.2	26.5	-3.8	1.7	0.9	4.0	24.7	1.4	7.7	11.2	25.7	54.0	0.0
1976	1.1	2.5	-7.1	1.7	4.2	-9.1	2.3	-0.5	0.2	-14.9	2.9	0.4	19.0
1977	8.0	8.3	16.5	1.6	11.0	10.3	2.4	27.9	9.5	-6.0	2.8	-0.3	20.1
1978	12.8	21.5	6.2	3.9	13.1	8.6	15.7	-2.0	17.0	-2.6	0.6	7.1	9.9
1979	26.3	38.4	10.5	11.9	26.9	13.0	19.1	5.3	32.8	18.9	6.8	15.5	-2.2
1980	7.8	2.7	18.3	14.1	18.3	20.7	38.1	16.4	17.6	8.7	8.9	11.0	-16.0
1981	12.2	13.7	9.7	9.3	12.5	0.6	19.3	17.3	8.3	8.4	8.7	7.8	0.0
1982	3.5	3.3	-3.0	3.2	5.1	-1.0	6.1	0.9	-2.9	23.5	7.8	13.7	0.0
1983	22.0	27.2	7.9	11.5	17.0	4.5	18.5	9.7	14.5	27.2	23.9	84.3	0.0
1984	28.3	18.6	56.6	31.7	54.6	41.4	43.2	54.0	35.3	72.8	39.3	29.8	6.2
February	0.6	0.9	0.0	-0.2	1.7	0.0	1.9	0.1	0.1	2.4	0.0	0.0	0.0
March	6.0	9.5	0.2	1.6	3.4	0.0	4.2	5.7	7.2	3.2	-2.1	0.0	0.0
April	-1.3	-3.3	0.0	3.0	0.2	0.0	0.0	1.0	1.6	4.2	0.1	0.0	0.0
May	6.7	2.5	32.0	4.6	18.4	29.3	0.0	19.4	2.3	2.4	15.3	4.8	0.0
June	1.3	-1.3	0.1	3.3	8.2	0.0	7.0	1.4	7.7	38.7	1.8	0.0	0.0
July	9.2	7.0	24.7	7.3	7.7	16.7	0.0	30.8	8.6	1.0	8.5	11.0	0.0
August	-1.5	-4.4	0.2	2.2	1.3	0.0	0.0	2.8	-0.3	0.0	1.1	0.0	5.0
September	3.9	6.4	8.1	1.2	-0.9	0.0	0.4	-2.4	3.4	0.0	5.3	0.0	4.7
October	12.5	23.4	0.0	1.2	6.3	0.8	0.0	-4.5	3.5	0.0	1.9	0.0	4.5
November	2.6	1.6	0.2	8.6	1.7	2.2	13.5	0.5	0.0	0.0	2.0	0.0	4.4
December	1.0	1.1	0.1	2.8	0.0	0.0	0.0	-0.1	-0.8	0.0	0.0	0.0	4.3
1985	23.4	12.9	36.0	39.9	42.5	22.4	22.3	23.5	28.0	68.3	31.0	40.7	22.4
January	5.5	8.3	2.2	4.0	-1.4	-5.8	-0.2	3.3	-0.5	27.6	-0.3	-12.4	4.0
February	-7.6	-14.7	0.0	4.1	0.4	0.0	0.9	4.4	0.7	0.0	-2.9	0.0	0.0
March	-4.4	-9.0	-0.7	1.1	3.0	2.2	0.2	-2.2	-0.4	-0.4	0.4	2.8	0.0
April	2.1	4.0	0.0	0.4	0.1	0.0	-0.3	0.0	0.5	0.0	0.8	0.0	0.0
May	-2.9	-8.5	0.2	2.1	8.6	2.2	1.1	-2.1	6.1	0.6	4.6	13.2	0.0
June	-2.3	-5.6	0.0	0.8	2.0	0.0	0.0	0.9	0.3	0.0	0.4	0.0	0.0
July	4.1	0.3	12.9	2.8	1.1	8.3	4.4	0.8	16.4	17.7	15.0	67.6	0.0
August	4.3	2.6	0.0	4.3	13.3	17.9	1.3	1.9	3.6	3.2	7.6	10.3	0.0
September	9.2	18.4	-0.4	2.3	4.0	-13.7	0.0	0.4	-1.0	10.6	1.3	0.0	0.0
October	3.4	7.0	0.2	2.9	0.0	0.0	-3.1	0.1	0.3	-0.9	-1.3	-15.9	0.0
November	-2.3	-3.2	-0.2	-6.3	7.4	6.2	17.0	-0.8	-8.1	-3.8	-2.3	-0.2	0.0
December	7.5	15.0	0.0	1.4	0.0	-15.8	-0.3	0.0	-2.6	0.0	-0.3	0.0	0.0
1986	45.1	64.0	40.3	24.1	45.1	46.6	42.1	6.2	29.9	8.9	24.9	15.8	1.1
January	14.3	25.0	-0.7	3.0	3.8	15.1	9.6	5.0	3.8	-0.2	3.3	-3.1	0.0
February	5.4	7.7	0.3	2.8	3.9	1.8	-0.2	-0.7	12.8	2.0	1.9	0.0	0.0
March	2.9	1.3	4.5	8.2	4.2	4.2	15.8	-0.3	5.1	-0.2	4.3	0.0	0.0
April	-3.7	-6.7	6.8	-1.7	1.6	3.4	0.0	0.0	1.9	0.0	-0.8	0.0	0.0
May	3.4	4.9	0.5	2.3	3.6	3.9	0.0	0.3	2.1	0.0	-1.1	0.0	0.0
June	-1.0	-3.0	0.1	3.7	1.4	0.0	0.0	0.0	1.9	0.0	-0.5	0.0	0.0
July	1.3	-0.3	30.9	1.1	3.2	25.7	0.1	0.5	4.0	-10.1	-1.5	0.0	0.0
August	3.1	3.2	4.9	-0.9	2.4	7.7	0.0	-0.3	-0.5	0.0	15.9	0.0	0.0
September	7.7	12.6	0.9	5.0	0.2	3.6	0.1	0.0	0.2	0.0	-0.5	0.0	0.0
October	0.0	-0.6	4.4	1.6	0.0	5.7	0.0	0.3	-0.4	0.0	-1.2	0.0	0.0
November	-5.0	-8.9	4.4	-2.3	-0.2	4.9	0.5	1.0	0.0	0.0	2.3	-2.9	0.0
December	-1.9	-5.1	0.0	2.8	0.0	0.0	0.0	9.8	-0.5	0.0	0.0	0.0	13.5

Source: Table 7.2.
16-Oct-91

Table 7.3: PARAGUAY - WHOLESALE PRICE INDEX 1972-1991
(Growth Rates)

continued

	General Index	Agriculture and Hunting	Forestry	Food Beverages & Tobacco	Textiles and Leather	Wood Products	Paper and Printing	Chemicals	Nonmetals Excluding Petroleum	Basic Metals	Machinery and Equipment	Other Manufacture	Electricity and Steam
1987	11.9	-1.1	68.9	23.3	28.1	68.4	21.9	23.6	18.2	-0.8	25.8	6.7	38.6
January	1.6	-0.9	11.6	-1.0	4.9	19.7	9.3	7.7	3.9	3.0	3.7	4.8	13.9
February	1.5	0.3	0.0	3.0	6.0	0.0	0.0	0.1	2.9	3.6	3.1	0.0	9.1
March	0.1	-1.1	0.4	0.2	1.5	0.0	6.6	0.8	0.6	0.0	2.4	14.9	0.0
April	-0.8	-2.2	0.1	1.1	0.0	0.9	0.0	0.0	0.7	0.0	1.2	0.0	0.0
May	0.9	-4.4	20.3	9.3	4.1	5.3	0.0	0.8	2.4	1.7	2.4	-6.8	0.0
June	2.1	2.8	0.0	1.9	3.8	0.0	0.0	0.4	1.1	0.0	0.3	0.0	0.0
July	1.5	1.2	6.8	0.6	0.7	4.1	0.0	4.4	3.5	-9.0	3.1	-5.0	0.0
August	-0.1	-0.3	1.0	-0.3	0.1	0.0	0.0	0.4	-0.2	0.1	0.1	0.0	0.0
September	7.3	12.6	0.3	3.8	4.2	2.4	0.5	0.7	1.6	2.8	2.4	0.0	0.0
October	6.9	12.0	0.0	4.1	2.2	0.0	0.5	0.3	-0.4	2.6	1.7	0.0	0.0
November	3.6	5.8	0.2	1.3	0.5	1.5	14.4	-0.8	1.8	1.9	3.4	-1.7	0.0
December	-5.9	-11.4	0.0	1.0	0.0	0.0	-0.2	0.1	0.1	-0.6	-0.3	0.0	0.0
1988	27.0	35.3	11.2	23.8	15.3	9.3	24.0	13.5	34.3	7.4	26.0	4.2	15.2
January	3.1	2.9	0.0	2.1	1.1	-3.2	1.8	7.7	7.0	1.4	9.2	13.0	0.0
February	3.3	5.6	0.2	2.0	0.5	0.0	0.0	0.3	2.7	0.0	0.9	0.0	0.0
March	3.4	5.8	0.1	2.0	1.3	0.9	-0.1	1.5	1.8	1.3	-0.4	-6.0	0.0
April	-2.3	-6.2	0.0	1.2	2.5	1.1	1.9	0.2	4.9	0.0	5.0	0.0	0.0
May	-0.6	-1.9	0.2	0.7	0.1	0.0	0.0	0.7	8.2	0.0	0.7	0.0	0.0
June	-4.8	-10.9	0.1	1.4	1.0	0.0	0.1	-0.2	2.4	0.0	0.6	6.3	9.5
July	5.9	12.8	0.1	-1.2	0.8	5.7	2.3	1.8	4.5	0.3	-1.0	-5.9	5.0
August	7.1	14.2	0.0	0.4	0.0	0.0	7.8	0.6	0.0	0.0	0.1	1.8	4.5
September	10.6	19.1	-0.1	3.0	0.3	0.2	0.0	0.2	0.5	9.4	1.4	-1.8	4.4
October	2.9	4.4	1.0	2.9	0.2	3.0	0.2	-3.9	-1.2	0.0	0.8	1.8	4.2
November	10.4	16.5	0.1	4.1	0.0	2.0	7.0	0.6	2.0	1.6	3.0	0.0	4.0
December	-7.9	-13.3	0.0	0.1	0.0	0.0	0.1	0.3	1.6	0.6	0.3	0.0	0.0
1989	21.9	18.1	11.0	27.7	19.2	27.0	20.5	35.5	36.6	40.1	26.9	28.7	73.9
January	0.6	-2.1	0.6	4.8	1.6	0.7	1.4	1.2	3.9	6.3	7.4	26.9	10.0
February	-0.7	-1.9	0.0	0.3	0.1	0.0	-0.1	1.1	1.6	0.0	3.2	0.0	9.1
March	2.4	3.5	0.0	-0.2	3.3	6.3	-0.6	1.9	-0.3	5.2	0.7	0.0	8.3
April	-0.7	-1.4	0.2	-0.1	0.1	0.0	0.1	0.5	0.7	0.0	0.7	0.0	0.0
May	0.9	-4.1	4.1	7.4	3.6	8.6	1.5	26.4	14.5	2.0	-0.3	0.0	0.0
June	-2.3	-8.1	2.2	1.3	5.6	3.3	1.6	7.6	9.6	5.6	5.0	-9.6	23.9
July	-2.2	-7.6	5.5	1.7	2.1	0.9	3.8	0.3	2.7	4.8	7.3	2.6	1.0
August	6.5	10.0	0.0	6.6	0.1	0.0	3.6	1.9	0.0	9.9	4.8	3.9	0.0
September	3.8	4.0	0.0	0.5	11.0	3.9	0.4	13.6	-1.3	7.5	0.1	3.9	0.0
October	3.8	2.7	5.1	11.3	1.8	3.4	-0.7	-0.8	0.0	3.4	0.3	4.8	0.0
November	6.0	14.4	0.1	-4.4	0.1	1.1	3.9	0.0	-0.1	0.5	-0.2	-1.6	0.0
December	5.9	12.0	6.4	-4.6	0.2	4.8	2.0	-0.2	0.1	0.5	4.0	-0.8	0.0
1990	55.1	76.8	36.3									19.2	27.3
January	4.4	7.1	-2.8									1.6	0.0
February	2.7	0.3	2.0									6.0	9.5
March	6.2	9.6	6.6									2.0	0.5
April	3.2	3.5	4.3									2.6	0.0
May	-2.6	-4.9	7.4									-0.4	0.0
June	7.4	10.6	4.7									3.9	0.0
July	7.4	11.8	0.0									2.0	0.0
August	0.4	0.5	-0.4									0.3	0.0
September	4.8	5.1	7.5									4.0	0.0
October	4.9	5.0	0.2									6.2	19.0
November	16.6	27.5	-3.5									1.6	0.8
December	-1.6	-2.9	-4.3									1.0	0.8
1991													
January	2.0	3.2	1.5									-0.1	0.0
February	4.3	5.5	0.3									2.3	0.0
March	2.1	3.5	-0.3									-0.2	0.0
April	-0.2	-0.8	2.1									0.9	0.0
May	-4.4	-10.3	0.3									7.1	0.0

16-Oct-91

Table 7.4: PARAGUAY - WAGE INDEX, 1969-1990
(1980 = 100)

	Manufacturing	Construction	Utilities and Sanitation	Commerce	Transport & Communication	Hotel and Domestic Services	General Index
Weights b/	57.9	44.0	8.1	0.6	27.4	1.6	100.0
1969	33.8	31.3	38.6	33.3	33.9	32.1	34.0
1970	35.2	31.4	38.6	33.3	34.8	33.7	35.1
1971	36.8	34.3	39.1	34.4	35.4	37.0	36.5
1972	38.4	34.3	39.1	35.2	37.2	37.2	37.9
1973	42.7	40.4	41.5	38.2	38.4	42.4	41.3
1974	51.3	51.0	49.2	51.5	42.9	52.2	48.8
1975	52.0	51.0	53.8	56.0	49.4	52.2	51.4
1976	57.1	58.5	59.4	56.5	54.7	52.2	56.6
1977	59.0	66.5	65.5	57.0	56.5	52.5	59.0
1978	68.1	74.5	72.6	63.3	64.5	60.2	67.6
1979	80.2	84.9	81.4	70.1	82.7	75.8	81.1
1980	100.0	100.0	100.0	100.0	100.0	100.0	100.0
1981	120.3	119.3	120.0	124.5	119.4	121.5	120.0
1982	122.7	119.9	128.8	126.7	128.2	121.6	124.6
1983	128.5	120.5	145.1	134.6	135.8	130.5	131.4
1984	153.6	148.4	162.2	163.9	146.5	171.6	152.4
1985	185.9	193.6	220.9	202.1	175.4	231.1	186.8
1986	238.2	257.7	268.1	268.5	211.8	304.1	235.4
1987	327.3	379.1	375.2	382.6	278.7	433.8	322.1
1988	443.2	519.6	463.1	528.0	353.8	589.3	426.8
1989	555.2	650.2	695.3	652.0	534.1	735.2	567.8
1990	686.5	788.9	819.3	744.7	663.8	867.8	698.3

a/ Wages paid to employees directly involved in production. Management is excluded.
b/ Based on the value of wages paid in June 1969.
Source: Central Bank of Paraguay.
16-Oct-91

Distributors of World Bank Publications

ARGENTINA
Carlos Hirsch, SRL
Galeria Guemes
Florida 165, 4th Floor-Ofc. 453/465
1333 Buenos Aires

AUSTRALIA, PAPUA NEW GUINEA, FIJI, SOLOMON ISLANDS, VANUATU, AND WESTERN SAMOA
D.A. Books & Journals
648 Whitehorse Road
Mitcham 3132
Victoria

AUSTRIA
Gerold and Co.
Graben 31
A-1011 Wien

BANGLADESH
Micro Industries Development
  Assistance Society (MIDAS)
House 5, Road 16
Dhanmondi R/Area
Dhaka 1209

Branch offices:
156, Nur Ahmed Sarak
Chittagong 4000

76, K.D.A. Avenue
Kulna 9100

BELGIUM
Jean De Lannoy
Av. du Roi 202
1060 Brussels

CANADA
Le Diffuseur
C.P. 85, 1501B rue Ampère
Boucherville, Québec
J4B 5E6

CHINA
China Financial & Economic
  Publishing House
8, Da Fo Si Dong Jie
Beijing

COLOMBIA
Infoenlace Ltda.
Apartado Aereo 34270
Bogota D.E.

COTE D'IVOIRE
Centre d'Edition et de Diffusion
  Africaines (CEDA)
04 B.P. 541
Abidjan 04 Plateau

CYPRUS
Cyprus College Bookstore
6, Diogenes Street, Engomi
P.O. Box 2006
Nicosia

DENMARK
SamfundsLitteratur
Rosenoerns Allé 11
DK-1970 Frederiksberg C

DOMINICAN REPUBLIC
Editora Taller, C. por A.
Restauración e Isabel la Católica 309
Apartado de Correos 2190 Z-1
Santo Domingo

EGYPT, ARAB REPUBLIC OF
Al Ahram
Al Galaa Street
Cairo

The Middle East Observer
41, Sherif Street
Cairo

EL SALVADOR
Fusades
Alam Dr. Manuel Enrique Araujo #3530
Edificio SISA, ler. Piso
San Salvador 011

FINLAND
Akateeminen Kirjakauppa
P.O. Box 128
SF-00101 Helsinki 10

FRANCE
World Bank Publications
66, avenue d'Iéna
75116 Paris

GERMANY
UNO-Verlag
Poppelsdorfer Allee 55
D-5300 Bonn 1

GUATEMALA
Librerias Piedra Santa
5a. Calle 7-55
Zona 1
Guatemala City

HONG KONG, MACAO
Asia 2000 Ltd.
46-48 Wyndham Street
Winning Centre
2nd Floor
Central Hong Kong

INDIA
Allied Publishers Private Ltd.
751 Mount Road
Madras - 600 002

Branch offices:
15 J.N. Heredia Marg
Ballard Estate
Bombay - 400 038

13/14 Asaf Ali Road
New Delhi - 110 002

17 Chittaranjan Avenue
Calcutta - 700 072

Jayadeva Hostel Building
5th Main Road Gandhinagar
Bangalore - 560 009

3-5-1129 Kachiguda Cross Road
Hyderabad - 500 027

Prarthana Flats, 2nd Floor
Near Thakore Baug, Navrangpura
Ahmedabad - 380 009

Patiala House
16-A Ashok Marg
Lucknow - 226 001

Central Bazaar Road
60 Bajaj Nagar
Nagpur 440010

INDONESIA
Pt. Indira Limited
Jl. Sam Ratulangi 37
P.O. Box 181
Jakarta Pusat

ISRAEL
Yozmot Literature Ltd.
P.O. Box 56055
Tel Aviv 61560
Israel

ITALY
Licosa Commissionaria Sansoni SPA
Via Duca Di Calabria, 1/1
Casella Postale 552
50125 Firenze

JAPAN
Eastern Book Service
Hongo 3-Chome, Bunkyo-ku 113
Tokyo

KENYA
Africa Book Service (E.A.) Ltd.
Quaran House, Mfangano Street
P.O. Box 45245
Nairobi

KOREA, REPUBLIC OF
Pan Korea Book Corporation
P.O. Box 101, Kwangwhamun
Seoul

MALAYSIA
University of Malaya Cooperative
  Bookshop, Limited
P.O. Box 1127, Jalan Pantai Baru
59700 Kuala Lumpur

MEXICO
INFOTEC
Apartado Postal 22-860
14060 Tlalpan, Mexico D.F.

NETHERLANDS
De Lindeboom/InOr-Publikaties
P.O. Box 202
7480 AE Haaksbergen

NEW ZEALAND
EBSCO NZ Ltd.
Private Mail Bag 99914
New Market
Auckland

NIGERIA
University Press Limited
Three Crowns Building Jericho
Private Mail Bag 5095
Ibadan

NORWAY
Narvesen Information Center
Book Department
P.O. Box 6125 Etterstad
N-0602 Oslo 6

PAKISTAN
Mirza Book Agency
65, Shahrah-e-Quaid-e-Azam
P.O. Box No. 729
Lahore 54000

PERU
Editorial Desarrollo SA
Apartado 3824
Lima 1

PHILIPPINES
International Book Center
Fifth Floor, Filipinas Life Building
Ayala Avenue, Makati
Metro Manila

POLAND
ORPAN
Palac Kultury i Nauki
00-901 Warzawa

PORTUGAL
Livraria Portugal
Rua Do Carmo 70-74
1200 Lisbon

SAUDI ARABIA, QATAR
Jarir Book Store
P.O. Box 3196
Riyadh 11471

SINGAPORE, TAIWAN, MYANMAR,BRUNEI
Information Publications
  Private, Ltd.
02-06 1st Fl., Pei-Fu Industrial
  Bldg.
24 New Industrial Road
Singapore 1953

SOUTH AFRICA, BOTSWANA
For single titles:
Oxford University Press
  Southern Africa
P.O. Box 1141
Cape Town 8000

For subscription orders:
International Subscription Service
P.O. Box 41095
Craighall
Johannesburg 2024

SPAIN
Mundi-Prensa Libros, S.A.
Castello 37
28001 Madrid

Librería Internacional AEDOS
Consell de Cent, 391
08009 Barcelona

SRI LANKA AND THE MALDIVES
Lake House Bookshop
P.O. Box 244
100, Sir Chittampalam A.
  Gardiner Mawatha
Colombo 2

SWEDEN
For single titles:
Fritzes Fackboksforetaget
Regeringsgatan 12, Box 16356
S-103 27 Stockholm

For subscription orders:
Wennergren-Williams AB
Box 30004
S-104 25 Stockholm

SWITZERLAND
For single titles:
Librairie Payot
1, rue de Bourg
CH 1002 Lausanne

For subscription orders:
Librairie Payot
Service des Abonnements
Case postale 3312
CH 1002 Lausanne

TANZANIA
Oxford University Press
P.O. Box 5299
Maktaba Road
Dar es Salaam

THAILAND
Central Department Store
306 Silom Road
Bangkok

TRINIDAD & TOBAGO, ANTIGUA BARBUDA, BARBADOS, DOMINICA, GRENADA, GUYANA, JAMAICA, MONTSERRAT, ST. KITTS & NEVIS, ST. LUCIA, ST. VINCENT & GRENADINES
Systematics Studies Unit
#9 Watts Street
Curepe
Trinidad, West Indies

UNITED KINGDOM
Microinfo Ltd.
P.O. Box 3
Alton, Hampshire GU34 2PG
England

VENEZUELA
Libreria del Este
Aptdo. 60.337
Caracas 1060-A

MAP SECTION